7th International Conference on Traffic and Logistic Engineering (ICTLE 2019)

MATEC Web of Conferences Volume 296 (2019)

Paris, France
21 – 23 August 2019

Editors:

El-Said Mamdouh Mahmoud Zahran
Roberto Montemanni

ISBN: 978-1-5108-9776-2

Printed from e-media with permission by:

Curran Associates, Inc.
57 Morehouse Lane
Red Hook, NY 12571

Some format issues inherent in the e-media version may also appear in this print version.

For additional information, please contact EDP Sciences – Web of Conferences
at the address below.

EDP Sciences – Web of Conferences
17, Avenue du Hoggar
Parc d'Activité de Courtabœuf
BP 112
F-91944 Les Ulis Cedex A
France

Phone: +33 (0) 1 69 18 75 75
Fax: +33 (0) 1 69 28 84 91

contact-edps@webofconferences.org

Additional copies of this publication are available from:

Curran Associates, Inc.
57 Morehouse Lane
Red Hook, NY 12571 USA
Phone: 845-758-0400
Fax: 845-758-2633
Email: curran@proceedings.com
Web: www.proceedings.com

TABLE OF CONTENTS

Preface

2019 7th International Conference on Traffic and Logistic Engineering (ICTLE 2019) was held in Paris, France during the period August 21st-23rd, 2019. The shared event provided a forum for accessing up-to-date knowledge, both from the industrial and the academic viewpoints.

The proceedings include both basic and applied research papers. All of the manuscripts were peer-reviewed by the conference committee members and by external international reviewers, leading to a high quality selection of great relevance for topics related to Logistic, Transportation Engineering and Industrial Engineering.

We would like to take this opportunity to thank the many people that contributed to the success of the event. In particular, the keynote speakers Prof. Khair S. Jadaan from University of Jordan, Jordan and Prof. Jagienka Rześny Cieplińska from WSB University in Gdansk, Poland.

We would like also to thank the committee members for their hard work in making this conference possible. Finally, many thanks to the reviewers for their excellent job to maintain the high academic quality of this conference.

Final thanks go to all authors and participants at ICTLE 2019 for helping to make it a successful event.

Conference Chair
Roberto Montemanni
University of Modena and Reggio Emilia, Italy

Statement of Peer review

In submitting conference proceedings to *Web of Conferences*, the editors of the proceedings certify to the Publisher that

1. They adhere to its **Policy on Publishing Integrity** in order to safeguard good scientific practice in publishing.
2. All articles have been subjected to peer review administered by the proceedings editors.
3. Reviews have been conducted by expert referees, who have been requested to provide unbiased and constructive comments aimed, whenever possible, at improving the work.
4. Proceedings editors have taken all reasonable steps to ensure the quality of the materials they publish and their decision to accept or reject a paper for publication has been based only on the merits of the work and the relevance to the journal.

Title, date and place of the conference

2019 7th International Conference on Traffic and Logistic Engineering (ICTLE 2019)

August 21-23, 2019

Paris, France

Proceedings editor(s):

El-Said Mamdouh Mahmoud Zahran

Roberto Montemanni

Date and editor's signature

September 28, 2019

September 29, 2019

Analysis of Black Spot of Traffic Accident in Wuhan Port

Feng Xia[1,2,a], Ya-dong Yang [2]

[1]*Hubei Provincial Key Laboratory of Inland navigation technique Wuhan 430063,China*
[2]*School of Navigation, Wuhan University of Technology, Wuhan 430063, China*

Abstract. The proposal to build the "Yangtze River Golden Waterway" strategy has enabled the shipbuilding volume of the Yangtze River trunk to grow rapidly. With the increase of ship traffic flow, the Yangtze River water traffic safety is facing severe challenges. This paper draws on the road traffic black spots and existing research methods of water traffic accidents and the definition of water traffic black spots. The DBSCAN algorithm and the accident-level weighted average method are used to establish the blackpoint identification model for waterborne traffic. The model is applied to the Yangtze River mid-stream Wuhan Port 2013-2017 water traffic safety accident as an example. Ten black spots and corresponding boundaries are identified. The black point analysis method used in this paper can effectively quantitatively analyze the spatial distribution and characteristics of water traffic black spots, which can provide a new theoretical basis for ensuring the safety of water traffic.

1 Preface

Waterway transportation is the main mode of transportation for bulk commodities. It plays an important role in the economic development at home and abroad. The safe navigation of ships has an important impact on the sustainable development of the shipping industry. Frequent waterborne traffic accidents have caused great losses to the country. To a certain extent, it has hindered the sustainable development of the port. In order to ensure the safety of the navigation environment of the ship, it is necessary to analyze and study the black spots of the water traffic in order to provide reference for the water traffic safety management of the maritime authorities and the port dispatching department, and also provide scientific and reasonable guidance for the ship driving personnel.

The research on black spots in water traffic is still in its infancy, and relevant scholars have used fuzzy comprehensive evaluation and hierarchical clustering methods to study the black spots of water traffic accidents. For example, Xu Yanmin et al[1] used the fuzzy comprehensive evaluation method to evaluate the ship collision risk, and identified the black point of the ship collision accident according to the hazard index identification standard and risk evaluation result of the black spot; Du Shanshan et al.[2] The class algorithm establishes the blackpoint identification model for water traffic, and studies the traffic black spots in the Zhangjiagang section of the Yangtze River. Domestic and foreign scholars have done a lot of research on road traffic black spots[3-7]. Bao Haitao[3] used DBSCAN-based analysis method to cluster point sets in space, thus achieving intelligent troubleshooting of traffic accidents. Zhang Man[4] combined with the theory of rough set and inclusion degree, proposed a property selection metric for constructing decision tree—including the importance degree. This method improves the clustering effect well. Michal et al.[5] combined the kernel density estimation with the clustering algorithm to identify black spots and identify the location of road black spots. Liu Anye et al[6] proposed the use of artificial neural network BP algorithm to identify the multiple points of road traffic accidents, enriching the basic theory of traffic accident multiple points identification. Saccomanno et al.[7] first proposed using Poisson regression and empirical Bayesian models to identify black spots in 2001. In 1974 Dunn[8] first proposed the ISODATA algorithm using fuzzy mathematics (Iterative Self Organizing Data Analysis Techniques Algorithm), by continually modifying the position of the condensation point to classify.

There is a certain similarity between road traffic black spots and water traffic black spots research, which can introduce the related research ideas of road traffic accidents into the study of water traffic accidents[9-10].

Therefore, this paper draws on the definition and identification method of road traffic black spots, based on the existing water accident-prone areas, defines the black spots of water traffic accidents, and proposes a water traffic black spot recognition model based on DBSCAN algorithm; using DBSCAN algorithm to analyze water Traffic accidents, realize reasonable and effective clustering of a large number of water accident data; combine the spatial distribution of water traffic accidents with the accident level, identify the black spots of water traffic accidents and mark the boundaries by studying the black spot identification of water traffic accidents, with a

[a] Corresponding author: 2998609852@qq.com

view to Quantitative identification of water traffic black spots to prevent only staying in qualitative analysis, in order to obtain more objective and accurate identification results, to provide reference for ship navigation safety.

2 Definition of water traffic black spots

The black spot of a traffic accident is the result of the cumulative distribution of traffic accidents within a certain period of time. According to the spatial distribution characteristics of water traffic accidents, the characteristics of accident levels and the definition of road traffic black spots [11-12], water traffic black spots can be defined as: The number or characteristic concentration of water traffic accidents in a certain period of time and specific waters In this case, the water area has an accident level higher than a certain limit. This paper fully considers the spatial distribution of water traffic accidents and the severity of accidents.

3 Water traffic black dot recognition model

This paper firstly uses the DBSCAN algorithm to cluster waterborne traffic accidents in space, and then uses the accident level comprehensive weighted average method to determine the blackpoint threshold. As a blackpoint recognition criterion, a waterborne traffic blackpoint recognition model is established. The advantage is that the DBSCAN algorithm is not only suitable for the processing of a large number of accident data, but also is classified by a gradual evolution method, and the method is more scientific; clustering the water traffic accidents from the space, combined with the difference of the accident levels, using the comprehensive weighted average of the accident levels The law distinguishes. The combination of these two methods not only takes into account the location of the accident space, but also the level of the accident.

3.1 The basic principle of the DBSCAN algorithm

The basic concept of the DBSCAN algorithm[3] is as follows:

1) Density: The density at any point in the space is the number of points contained in the area where the point is the center of the circle and the radius is E_{PS}

2) Neighborhood: The neighborhood of any point in the space is a set of points contained in the area where the point is the center of the circle and the radius of E_{PS} is recorded as $Neps(p)=\{q \in D|dist(p,q) \leqslant E_{PS}\}$ Here D is the database.

3) Core point and boundary point: The density of a point in space, if it is greater than or equal to a given threshold Minpts, it is called the core point; otherwise it is called the boundary point.

4) Direct density is reachable: point p is directly accessible from point q and satisfies $p \in Neps(q)$, Neps(q)>Minpts.

5) Density up: In database D, if there is a set of objects p_1, $p_2 \ldots p_n$ makes $p_1=q$, $p_n=p$ for any $p_i \in D(1 \leqslant i \leqslant n)$, p_{i+1} is from p_i about E_{PS} and Minpts are directly reachable, so p is called q from E_{PS}.and Minpts.

6) Density connected: There is a point o such that point p and point q are from o for E_{PS} and Minpts density, then point p and point q are connected to E_{PS} and Minpts density.

7) Class: The non-empty set C of the database D is a class, if and only if p, q, if $p \in c$, and from p density up to q, then $q \in c$; p, q, if $q \in c$, Then p and q are density connected.

8) Noise: Points in the database D that do not belong to any class are noise.

3.2 Implementation of DBSCAN algorithm

The implementation process of the DBSCAN algorithm is as follows:

1) Initialize the settings.

(1)Add a new attribute ClusterID for storing the clustering result in the original database D, initialize it to zero, and form a new data set D_1.

(2) Define TempList to temporarily store the point where the distance between two points is less than the radiuse.

(3) Set the parameters Minpts and E_{PS}.

2) Traverse the set D_1 and sequentially examine each point as a seed point. i=1, j=1, and Cluster=1.

(1) For any point p_i, if it's ClusterID=0, search for its E_{PS} neighborhood; if $Neps(p_i) \geqslant$ Minpts, then point p_i is the core point, let it's ClusterID=Cluster, and it's E_{PS} neighborhood All points included are stored in the TempList.

(2) Traverse the TempList and examine each point as a seed point. For any point q_j, if the cluster ID of q_j =0, search for its E_{PS} neighborhood. If $Neps(p_j) \geqslant$ Minpts, then q_j is also a core point. It is the direct density reachable point of point q_j, which belongs to the same class as p_i, so that it's ClusterID=Cluster, otherwise q_j is the boundary point, but it's ClusterID=Cluster. If q_j is the core point, point $o \in Neps(q_j)$, and o does not belong to TempList, then point o is stored in TempList, and finally point q_j is deleted from TempList.

(3) j=j+1, if TempList is not empty, step (2) is performed.

3) i=i+1, Cluster=Cluster+1, step (2) is performed until the data set D_1 is traversed.

4) Delete the search data set TempList.

At this point, the cluster ends, and only the search is needed for the data set to get the final result. The clustering result is recorded in the D_1 attribute database, and the point where the field ClusterID=0 is a noise point.

3.3 Water Traffic Black Spot Recognition Based on Accident Level Weighted Average Method

In combination with the "Water Traffic Accident Statistics Method", the classification of water traffic accidents is used. This paper uses the gradient assignment method commonly used in the study of water traffic safety to convert the severity of the accident into a ratio to distinguish the severity of different levels of accidents. Table 1. For the accident class p_i obtained by spatial clustering based on the DBSCAN algorithm, a comprehensive weighted method is used to determine the water traffic black spots. First, according to the coefficients in Table 1, the different levels of accidents in the accident class p_i are assigned and summed to obtain the blackness value Qi of the accident class p_i Secondly, the blackness values Q_i of all accidents in the entire jurisdiction are summed. Obtain the black value of the accident in the whole research area ΣQ;

Table1 Coefficient value of the different .grade accident

Serial number	Accident level	Weight
1	Small accident	1
2	General accident	2
3	Big accident	3
4	Major accident	4
5	Particularly serious accident	5

Finally, obtain the average value q of all blackness values of the accidents in the study area, and use this value as the judgment threshold of the black spot of the water traffic. If the blackness value of an accident class p_i is greater than the average value Q, it can be determined that the accident class p_i is a water traffic black spot.spot.

$$Q_i = n_{i1} \times W_1 + n_{i2} \times W_2 + n_{i3} \times W_3 + n_{i4} \times W_4 + n_{i5} \times W_5 \qquad (1)$$

$$\overline{Q} = \frac{\Sigma Q}{K} \qquad (2)$$

In the formula (1), Q_i is the blackness value of the accident class p_i; i is the accident class number, i= 1, 2, 3, ..., k; n_{i1} is the number of small accidents in the accident class; n_{i2} is the general accident in the accident class p_i The number of n_{i3} is the number of major accidents in the accident class p_i; n_{i4} is the number of major accidents in the accident class p_i; n_{i5} is the number of special accidents in the accident class p_i; W_1, W_2, W_3, W_4 and W_5 are the weights of the five accident levels in Table 1., and k in the formula (2) is the total number of accident classes in the whole study area.

3 Application examples

3.1 Overview of Wuhan Port

Wuhan Port is located in the middle reaches of the Yangtze River, the Zhuankou Port Area (mainly for car rolling), the Zuoling Port Area (mainly for chemical and dangerous goods handling), and the Yangshuo Port Area (mainly for container and bulk cargo handling) Hankou Port Area (mainly for tourist passenger transportation and leisure), Qingshan Port Area (mainly steel and ore), and Yangluo Port Area (mainly container and integrated logistics). Its navigation channels include Hankou Waterway, Wuqiao Waterway, and Qingshanjia Waterway.

From 2003 to 2017, there were 434 water traffic accidents in Wuhan Port, including 393 small accidents and 41 minor accidents. According to the statistics of the maritime affairs department, the above-mentioned 434 accidents occurred in the waters of Hankou Waterway, Wuqiao Waterway, Qingshan Clip Waterway, Tianxing Island Anchorage, and liquid cargo ship waiting for anchorage. Use Acr GIS visualization and other functions to assist in the spatial distribution of accidents. By importing the accident location coordinates, you can display the spatial distribution of the water accident on the map.

Fig.1 Real scene distribution of water accident in wuhan

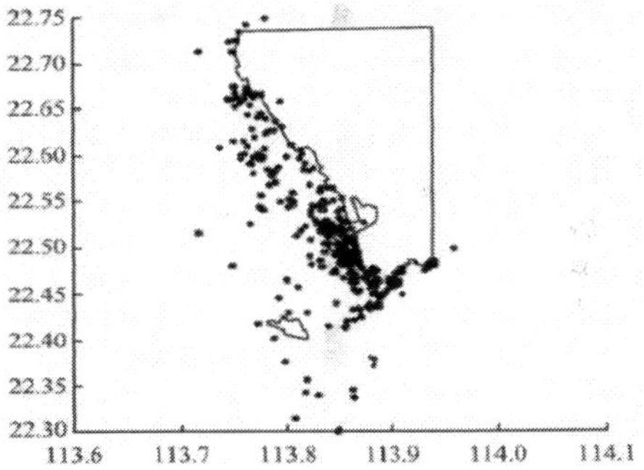

Fig.2 Distribution of water traffic accident in wuhan port

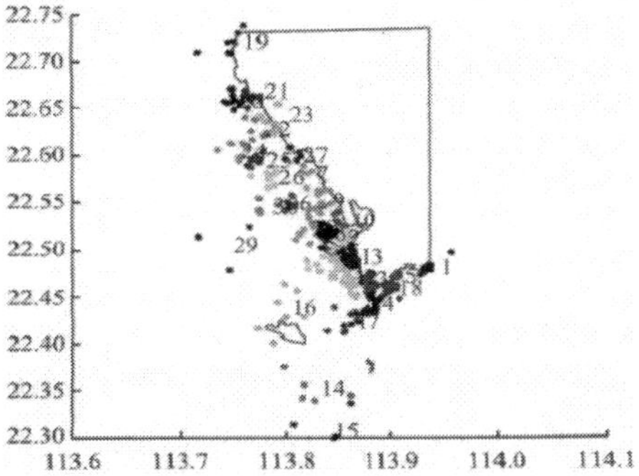

Fig.3 Clustering resuits of waterway traffic accident with 30 clusters

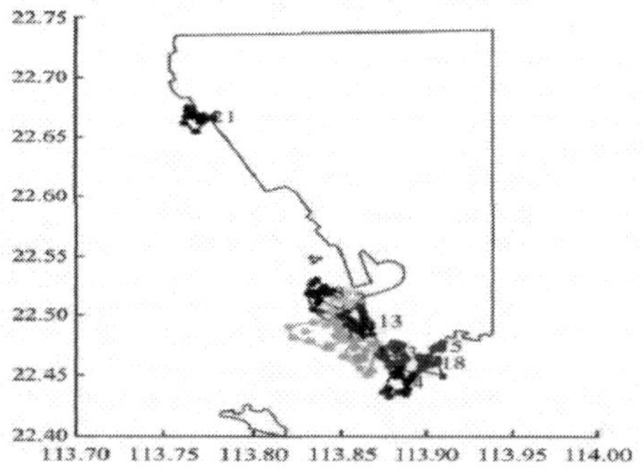

Fig.4 Position of 10 waterway traffic accident black-spots

The distribution of accidents in Wuhan Port is shown in Figures 1 and 2.

3.2 Identification of waterborne black spots in Wuhan Port waters

First, construct a geographic coordinate data matrix D of the accident in the study area. Secondly, using ArcGIS to import the coordinates of the accident location, the spatial distribution of the complete water accident can be displayed, as shown in Figure 2. Finally, the initial values of the control parameters Eps and Minpts are selected, and the accident clustering is obtained through the intelligent DBSCAN algorithm. This paper refers to a new method to determine Eps and Minpts adaptively by analyzing the statistical characteristics of data sets. This method is based on DBSCAN algorithm and proposes a new method to determine parameters Eps and Minpts adaptively based on distance, which is called I-DBSCAN. The main idea of this method is to select the Eps value and Minpts value according to the statistical characteristics of the data set itself, which is finally determined as Eps =5km and Minpts=10 by this method[13-14]. and 434 water traffic accidents in the study area are classified into 30 categories, as shown in Figure 4. According to equations (1) and (2), the blackness value and the accident blackness judgment threshold of each accident type are calculated, as shown in Table 2.

Table2 Risk degree of each accident cluster and threshold value of the accident black-spot

Clustering	Blackness value	Clustering	Blackness value
1	10	16	8
2	51	17	13
3	42	18	23
4	28	19	10
5	21	20	10
6	5	21	27
7	5	22	7
8	26	23	11
9	15	24	10
10	13	25	11
11	31	26	9
12	23	27	8
13	53	28	8
14	5	29	6
15	6	30	8
Threshold: 16.77			

Comparing the blackness value of each accident category and the blackout point threshold of the accident in Table 3, it can be seen that there are 10 accidental blackness values greater than the blackpoint threshold of the accident, so the 10 locations are determined as black spots of water traffic accidents, as shown in Figure 5. The 10 identified water traffic black spots are marked from the outside (in the order of extension from the river to the shore), which are 1 (category 4), 2 (category 18), 3 (category 5), 4 (Category 3), 5 (Category 12), 6 (Category 11), 7 (Category 13), 8 (Category 2), 9 (Category 8), 10 (Category 21), respectively Corresponding to Hankou waters (No. 1 black spot), Hankou waters (No. 2 black spot), Hankou waters (No. 3 black spot), Tianxing Island anchorage (No. 4 black spot), Qingshan clip water area (No. 5 black spot), Qingshan Clip waters (black spot 6), Qingshan clip waters (black spot 7), tanker waiting for anchorage (black point 8), Wuqiao waters (black point 9), Wuqiao waters (black point 10) The boundary range of these 10 black points (the boundary accident point number) is shown in Figure 5, respectively.

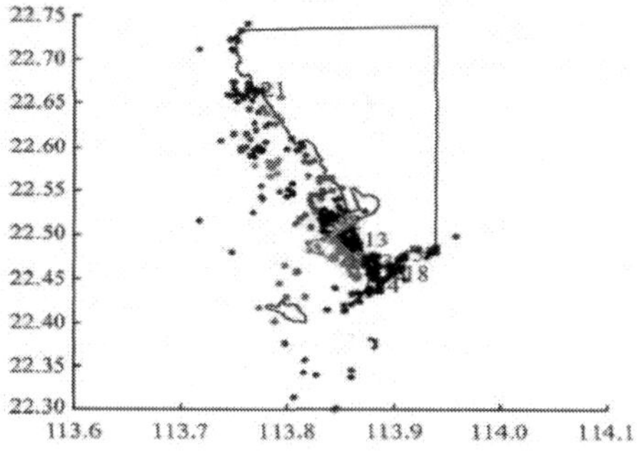

Fig.5 Boundary of 10 waterway traffic accident black-spots in wuhan port

Black spot No. 1: The area enclosed by the line of 103-374-255-244-375-373-103.
Black spot No. 2: The area enclosed by the line of 28-225-287-284-127-28.
Black spot No. 3: The area enclosed by the line of 31-292-69-90-222-136-294-31.
Black spot No. 4: The area enclosed by the line of 6-58-256-233-338-416-295-101-148-106-6.
Black spot No. 5: The area enclosed by the line of 12-163-381-56-164-12.
Black spot No. 6: The area enclosed by the line of 80-137-419-86-223-179-224-102-80.
Black spot No. 7: The area enclosed by the line of 20-327-158-321-182-411-313-226-414-20.
Black spot No. 8: The area enclosed by the line of 9-227-401-354-355-300-275-274-213-9.
Black spot No. 9: The area enclosed by the line of 7-176-420-368-394-196-112-7.
Black spot No. 10: The area enclosed by the line of 37-149-151-237-345-140-37.
According to the blackness value of each black point, the black point risk degree is from the highest to the lowest,

7th black point, 8th black point, 4th black point, 6th black point, 1st black point, 10th black point. , 9th black dot, 5th black dot, 2nd black dot, 3rd black dot.

The location identified by the maritime traffic black-spot recognition model is consistent with the maritime department's statistically concentrated multiple locations, and the location of the black-spot recognition results is more specific and the boundaries are more clear, which is conducive to water traffic supervision.

4 Conclusion

1) A black spot recognition model for water traffic is established. Using the DBSCAN algorithm and the accident level weighting method, the water traffic black spots are studied from the spatial distribution with the number of accidents and the accident level as indicators.

2) Based on the built model, a case study of Wuhan Port 2013-2017 traffic accident is carried out to identify the traffic black spots and corresponding black point boundaries of the water area, and determine the severity of each black point, which is consistent with the accident occurrence position. And the boundaries are clearer.

3) In the follow-up study, the research on the causes of black spots in water traffic accidents will be carried out based on the identified black spots of water traffic accidents.

References

1. XU Yanmin , TANG Chengang . Identifying gloomy spots of ship collision accident with grid theory [J] . Navigation of China, 2013, 36(4):72-75.

2. DU Shanshan , CHEN Houzhong Identification of waterway traffic black-spot [J] Journal of Dalian Maritime University, 2016, 42(2):46-5.

3. BAO Haitao. Evaluation of accident black spots on roads using clustering algorithm [D] . Changchun: Jilin University, 2005.

4. ZHANG Man . Research on clustering and classification algorithm based on rough set and inclusion degree [D] . Qingdao: Qingdao Technological University, 2015.

5. MICHAL B , RICHARD A , ZBYNEK J . Identification of hazardous road locations of traffic accidents by means of kernel density estimation and cluster significance evaluation [J] . Elsevier, 2013, 55(3) :265.

6. SACCOMANNO F , GROSSIR , GRECO D , et al . Identifying blackspots along highway SS107 in southern Italy using two models [J] . Journal of Transportation Engineering. 2001, 127(6) : 515-522.

7. DUNN J C . A fuzzy relative of the ISODATA process its use in delecting compact well-separated clusters [J] . Journal of Cybenetics, 1974, 3(3) : 32-57.

8. XIAO Shen , GUO Xiucheng , XU Jiandong . Application fuzzy clustering method in the highway traffic black-spot cause of formation analysis [J] . Journal of Transportation Systems Engineering and Information Technology, 2002, 2(3) : 40-43.

9. WANG Hai . Based on spatial analysis technology of traffic accident black point identification and cause analysis [D] . Beijing: Tsinghua University, 2014.

10. SANDHU H A , SINGH G , SISODIA M S , et al. Identification of black spots on highway with kernel density estimation method [J] . Journal of the Indian Society of Remote Sensing, 2016, 44 (3) :457-464 .

11. GEURTS K , WETS G , BRIJS T , et al . Identification and ranking of black spots: sensitivity analysis [J] . Transportation Research Record Journal of the Transportation Research Board, 2004, 1897(1) : 34-42.

12. ANDREAS G , KYRIACOS C M . Black spots identification through a Bayesian networks quantification of accident risk index [J] . Elsevier, 2013, 28(3) : 28-43.

13 . Jiang Hua, Li Jing, Yi Shenghe , et al . A new hybrid meth—od based onpartitioning—based DBSCAN and antclustering[J] Expen Systems with Applications , 2011, 38(8):9373-9381.

14 . Hong FangZhou,Peng Wang. DBSCAN algorithm for parameter adaptive determination method [J]. Journal of xi 'an university of technology,2012,28(3):289-292. DOI:10.3969/j.issn.1006-4710.2012.03.007.

Analysis of Key Issues on Man-Control to System-Control Leap in Autonomous Driving

Zongwei Liu[1,2], Hong Tan[1,2], Han Hao[1,2], Fuquan Zhao[1,2,a]

[1] *State Key Laboratory of Automotive Safety and Energy, Tsinghua University, Beijing 100084, China*
[2] *Tsinghua Automotive Strategy Research Institute, Tsinghua University, Beijing 100084, China*

Abstract. Autonomous driving is recognized as a global development direction and a major opportunity. The function and use of the vehicle has changed profoundly. The vehicle is gradually transformed from a simple transportation tool to a smart mobile space. The ultimate goal of autonomous driving is to achieve driverless driving. In the course of its development, man-control will gradually turn into to system-control. In other words, the transition from level 3 (L3) to level 4 (L4) is a fundamental leap. At present, the specific path to achieve this leap is not yet clear. Different companies have different and even opposite thinking and choices. In this study, the grading standard for autonomous driving was clearly explained, and the technical route selection of the company was analysed. Based on the analysis, the requirements of sensing, decision making, execution between the L3 and L4 were compared. Moreover, the key technical difficulties of L3 to L4 were clarified. In the end, suggestions on the commercialization of autonomous driving were given.

1 Introduction

A new round of scientific and technological revolution represented by cloud computing, big data, artificial intelligence is in the period of outbreak [1]. The technological revolution has brought about profound changes in the global automotive industry [2]. As the main core component of the traditional manufacturing industry, the automobile industry is entering an unprecedented period of change [3]. Traditional auto companies and technology companies are relying on multiple technology paths and multiple business models to achieve self-driving [4]. The functionality and application scenarios that automobiles can achieve are changing dramatically. Vehicles will gradually change from simple means of transportation to high-speed mobile office and living places, as well as information control terminals [5]. Driverless driving cars can achieve far richer functions and services than traditional cars.

Traditional car companies and technology companies are relying on a variety of technology paths, through a variety of business models to turn autonomous driving into reality. With kinds of advanced sensors, controllers and actuators, intelligent network automobile products with autonomous driving function have gradually entered the market [6]. In the practice of autonomous driving industry, some companies put forward the technical route of direct research and development of L4 products instead of L3 products. On the contrary, some auto companies still think it necessary to develop L3 products first. Different enterprises have very different thinking and choices. It's because the specific path to achieve the leap from L3 to L4 is not clear yet.

This paper is organized as follows. The next section describes the classification principle of autonomous driving and different choices of autonomous driving technology paths in current enterprises. Following that, technology paths between man-driven L3 and machine-driven L4 was compared. The subsequent section identified and analysed the key problems of leaping from L3 to L4. The final section provided suggestions on the commercialization of autonomous driving.

2 Core Concepts and Key Technology Paths

2.1 Core concepts of autonomous driving

The L3 is a key stage in the transition from man-control to system-control. According to the latest revision of SAE J3016(TM) by SAE International, autonomous driving is subdivided into six levels from L0 to L5 [7], as shown in Figure 1. The classification is based on the division of responsibility for performing steering and speed changes, the responsibility of monitoring the driving environment, the responsibility of dynamic driving task retreat and the driving system capability. Six levels correspond to the improvement of the degree of machine replacing man driving.

[a] Corresponding author: zhaofuquan@tsinghua.edu.cn

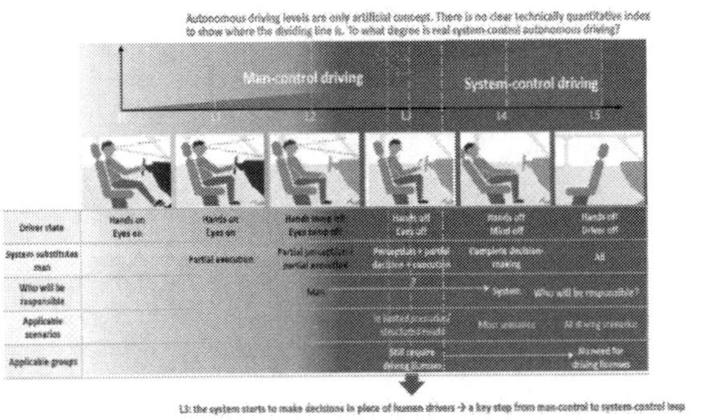

Figure 1. Classification principle of autonomous driving.

L3 self-driving allows human drivers to drive with their hands and eyes separated. But when the system makes a request, the human driver must respond appropriately. Machine replacements for human complete perception, execution and partial decision-making at the L3. However, the L4 autonomous driving can replace all decision making decisions and all driving operations. When the system issues a takeover request, the vehicle will ensure its own safety even if the human driver does not respond.

In fact, the classification is just a human concept. At which stage to achieve the transition from man-control to system-control, there is no recognized exact definition and quantitative technical indicators. As far as the definition of SAE is concerned, it is clear that the L3 level is a very critical step in achieving the leap. Because the machine has begun to replace people to make some decisions at this point. And the L4 level has already undoubtedly entered the stage of machine-led driving.

2.2 Different technology paths choices for L3 and L4 autonomous driving for enterprises

In the practice of autonomous driving, some companies have proposed to directly develop L4 products across L3. In contrast, other companies believe it is necessary to introduce L3 products first. For companies that choose to develop L4 directly, the considerations include user experience and security risks. From the perspective of user experience, some traditional car companies believe that L3 products have no commercial value. The reason is that the vehicle's irregular prompts and requests to take over the driving rights will cause confusion to the driver and even make it more tense than driving a traditional vehicle. Moreover, if the driver fails to take over within a limited time or can not take over the driving, L3 products will bring huge security risks. Indeed, this kind of unexpected situation cannot be completely avoided. In order to solve this problem fundamentally, L3 must be able to achieve a high level of driving. In other words, the autopilot does not send a request to the human driver in most cases. However, such L3 is very close to L4. Therefore, it is better to directly develop L4 products.

The company that advocates skipping L3 and heading straight to L4 is a radical. Its market advantage is that it is easier to get a larger share in emerging markets such as taxi services, automatic distribution services, and mobile travel services. Those who occupy market share first can give priority to building brand value, win consumer trust, and obtain possible natural monopoly. Its design and manufacturing advantages can be "one step in place". In another words, companies could focus on designing a centralized electronic control architecture for driverless driving at the beginning, instead of adding patches to the regional electronic control unit.

In contrast, most car companies represented by Audi still believe that L3 products have commercial value. It is believed that L3 products can solve user pain points under certain circumstances. For example, Audi has launched the L3 product. The A8L can achieve autonomous driving in scenarios to ensure safety such as low speed (60km/h or less) and less dangerous urban congestion roads. At the same time, sufficient time to take over is reserved on the interaction scheme. Use step-by-step alerts to optimize user experience. Considering that the recent launch of L4 products is very difficult, car companies make this technical route choice. Not only does it take time to do technical iterations and cost reductions, but there are also issues of consumer trust and acceptance. However, in order to seize market opportunities as soon as possible, the introduction of L3 products has become a good choice to increase the selling point of products.

The company that advocates a gradual evolution from L3 to L4 are conservative. The technological advantage is that the number of sensors required for L3 products is small. In addition, L3 products have low requirements for sensor performance, software complexity, computing power, redundant design and infrastructure. The presence of L3 also gives human drivers the opportunity to become familiar with handing over driving tasks to the system. However, the technical route pays more attention to the reliability of the technology. The technology should enable the automated driving system to handle any situation within 5 to 10 seconds necessary for the human driver to be aware of the need to take over and control. Because if a certain accident is caused, the negative public opinion will offset the gain brought by L3. More seriously, it also delays the time when L4 products enter the market.

In fact, the choice of two technology paths is reasonable. Specifically, whether to develop L4 directly or launch L3 first depends not only on the orientation and planning of the enterprise for its own development, but also on the essential difference between L3 and L4. More importantly, enterprises should have a comprehensive and accurate understanding of key issues on man-control to system-control.

3 Comparison and leaping difficulties

3.1 Comparison of L3 and L4 autonomous driving

Autonomous driving technology includes perception, decision-making and execution. Although there are some similarities or inheritances in these three aspects, there are still quite different between L3 and L4, as shown in Figure 2.

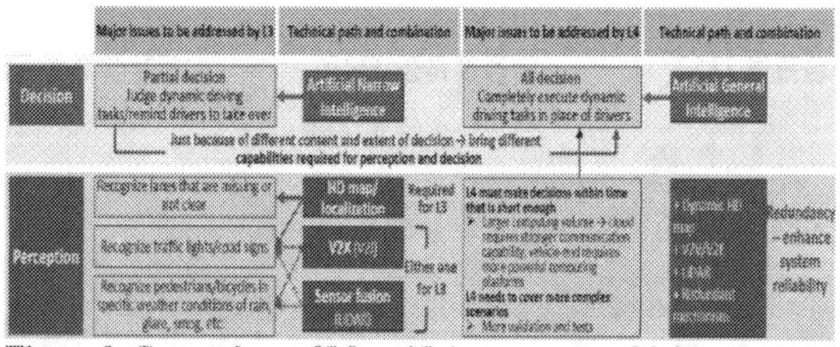

Figure 2. Comparison of L3 and L4 autonomous driving

From the perspective of decision making, applying general artificial intelligence to L3 can be sufficient to meet the needs of partial decision-making, including judging dynamic driving tasks and alerting human driver to take over. In contrast, L4 requires the application of strong artificial intelligence to completely replace people for all decisions and perform dynamic driving tasks. Due to the different content and degree of decision-making, the corresponding capabilities required for perception and decision-making are completely different.

From a perceptual perspective, L3 uses high-precision maps and positioning to identify missing and blurred lane lines. Identify traffic lights and road signs via V2X (car network) [8] or sensor fusion. And identifying pedestrians and bicycles under certain conditions such as rain, glare, and haze [9], [10]. For L4, dynamic high-precision maps, V2X and sensor fusion including Lidar are indispensable. Ensure that decisions can be made in a short enough time and cover more complex scenarios. In addition, redundant mechanisms are needed to enhance the reliability of the system.

From the perspective of implementation, the technical route of L3 and L4 autonomous driving is basically the same. Linear control of actuator is a necessary condition for autonomous driving above L3. On this basis, L4 enlarges the redundancy of actuator.

The choice of two technology paths is to develop L4 directly or release L3 first. The core of the controversy is whether L4 can be developed gradually through the accumulation and promotion of L3 technology. Based on the above analysis, L3 and L4 are not simple progressive relationships. This is not only a qualitative leap in technology, but also a quantitative leap in workload. In addition, it involves many issues outside the technology. Therefore, the premise of achieving L4 is to achieve a fundamental breakthrough. However, the development of L3 can indeed support L4 in some aspects, including the accumulation of common technologies such as sensors and algorithms.

3.2 Key difficulties in leaping from L3 to L4

3.2.1 decision-making

At present, the sensing and execution technological route of autonomous driving has been relatively clear, but the reliable decision-making technological route is still inconclusive. As environmental complexity and task complexity continue to increase, the decision-making techniques required for autonomous driving must also be upgraded. The environment goes from a controlled environment to a regular environment, and finally to an unpredictable environment. The driving task changes from the pre-set destination to the predicted destination, and finally to the destination dynamically. Environmental complexity and task complexity are gradually increasing. The decision-making techniques required for autonomous driving must also be upgraded. Decision-making techniques should respond to change from established programming methods based on rules to machine learning based on scene and conditional probabilities, and finally to deep learning based on neural networks with automatic learning capabilities. In other words, the transition from L3 to L4 needs to adapt to more complex tasks and environments, and the decision-making difficulty varies greatly.

Decision-making of L4 requires deep learning. There are two main types of deep learning methods currently applicable to autonomous driving, namely semantic abstraction and end-to-end learning. Semantic abstraction is more common in the choice of automotive companies. he principle is to decompose the automatic driving problem into several modules such as detecting vehicles, lanes, obstacles. And then expert expertise and experience are used to tag and annotate your data to train each module. The advantage of this method is the low rate of failure. Once the error occurs, the source can be traced back and the unknown situation of the burst can be handled. The disadvantage is that a lot of up-front preparation and programming is required during the training process. And a large amount of redundant data marks are generated.

And end-to-end deep learning is more concerned by IT companies. The principle is to directly learn the real driving data of human beings. The system is equivalent to a black box of input perception and output decision. The advantage of this method is that it does not require precise marking and annotation of scene data during training., and the system as a whole for self-learning and optimization. As training data increases, the system becomes more reliable. The disadvantage is that a higher level of artificial intelligence engineers are required to train the system with a larger amount of data. In essence, both of these deep learning methods are highly dependent on big data, and getting full and complete big data is far more difficult than imagined.

3.2.2 intuition

Human drivers can intuitively predict the behavior of other road traffic participants while driving. This plays an important role in the safety of drivers and others. However, it is very difficult for intelligent vehicles to understand human intentions and predict human behavior. This is also one of the fundamental challenges facing machine-driven driving. There is no recognized solution for this problem in the industry. There are three possible cracking methods. The first is to let the machine thoroughly understand human behavior and intuition, through in-depth research in brain science, behavioral science, and cognitive science. The second is to completely remove the human driver from the road traffic system. The focus of this problem is changed to the pre-judgment behavior of intelligent vehicles. The third is to let the machine "learn" the intuition through various algorithm upgrades, so that the machine is infinitely close to people. Obviously, these three paths are difficult to achieve in the short term.

3.2.3 Testing and verification

At present, the basic idea of using artificial intelligence technology to solve the problem of autonomous driving is relying on massive data. In order to solve problems accurately and effectively, it is necessary to ensure the completeness of the collected big data. The study pointed out that if the human driver's driving safety data is used as a reference, in order to prove that the active safety performance of the self-driving car exceeds man-control, the required test mileage may be as high as 5-6 million kilometres [11]. In addition, traditional vehicles only need to test the reliability of vehicle control and actuators, while intelligent vehicles also need to test the ability of perception, decision-making and related hardware and software. Correspondingly, it brings some new problems. For example, accidents that have not been recorded in the history of man-control also need to be avoided by autonomous driving vehicles. How to collect such scene data comprehensively? Another example is that intelligent vehicles may cause new traffic accidents. How to accumulate and verify such scenes? In addition, it is also necessary to consider how people's driving behaviour will change in the environment where people and machines work together.

The only way to solve the problem is to continuously increase the accuracy and proportion of virtual testing through scientific methods based on closed environment and open road measured data [12]. But the virtual test model does inevitably simplify the real situation. And the verification of its effectiveness is very difficult. Obviously, there is still a long way to go.

In addition to the above three major difficulties, autonomous driving has problems other than technology to face, including the problems about human trust in machine-driven vehicles, social ethical issues in which machines replace people making decisions [13], and laws and regulations that are contrary to current regulations. In the long run, these social problems will eventually be resolved with changes in consumer concepts, advances in technology, and improvements in regulations. But this takes time and must be a long-term process. For companies, the judgment of this time process is closely related to product and technology planning.

4 Suggestions on the commercialization

First of all, the fact is that L4 products that meet the needs of various scenarios do not have the possibility of mature mass production in the short term. Because L4 is fundamentally different from L3 in terms of development difficulty, technical cost, subject responsibility and social issues. In addition, there are still many problems that cannot be solved simply by relying on the car company. In other words, the most likely automatic production in the near future is the L3 product. Recently, some companies have expressed that they have been too optimistic about the progress of the autonomous driving development process. Therefore, if the company only layouts and develops L4 products, it is likely to lose the product selling point that is recently needed. Even enterprises will face the risk of mass production in the foreseeable future. Even if choosing the technical route to directly develop L4, companies should also develop L3 at the same time. This is not only a backup guarantee that the time for launching L4 products may lag, but many technical developments of L3 can also provide effective support for L4. In general, for companies with relatively limited strength and resources, it is more important to cut into L3 to achieve short-term gains, regarding the L4 technology as a long-term technical reserve.

Secondly, people do not need to be completely liberated during the driving process in the current stage. If the autonomous driving can effectively liberate the driver in certain scenarios for daily use, there is a good market prospect. Of course, the premise is to make sure the experience is good enough. For example, in a closed environment such as automatic parking in a parking lot, highway driving, and low-speed urban congestion conditions, L3 products are sufficient to meet consumer demand. This brings great convenience to consumers. Of course, the premise is that companies must make their L3 products good enough. In order to ensure the safety of man-machine switching during driving, it is necessary to make it as close as possible to L4. The bad driving experience caused by frequent requests for the driver to take over the vehicle should be considered. In this sense, there is still a lot of work to do for L3. In addition to continuing to increase investment in technology research, companies should also evaluate from the two dimensions of autonomous driving grading and driving scene grading. Enterprises should comprehensively analyze and systematically plan application scenarios suitable for different levels of autonomous driving products. Based on the difficulty of implementation, the expected industrialization time note for enterprises should be given.

Finally, in order to achieve sustainable development, enterprises must be based on the present and the future. For this reason, they must always be prepared. For autonomous driving, companies must balance the research reserve of technology with the mass production application of the product, and actively explore potential market opportunities corresponding to technology, as well as try innovative business model. In this sense, companies can completely customize the connotation of L3 and L4 products, and accelerate the promotion of autopilot products under common scenarios. At the same time, the current stage of car companies must consider the research reserve of L4 autonomous driving technology. One thing to emphasize is that the development of L4 should be effective in feeding back L3 autonomous driving technology products, rather than letting L4 technology and products develop in isolation. In specific practice, car companies should focus on strengthening cooperation and reserves in artificial intelligence, continuously accumulate and improve the database of autonomous driving application scenarios, and pay attention to and promote the development of new testing and verification methods. In specific practice, car companies should focus on strengthening cooperation and reserves in artificial intelligence. At the same time, the autonomous driving application scenario database needs to be continuously accumulated and improved. In addition, companies should pay attention to and promote the development of new testing and verification methods. Only in this way can the enterprise be invincible in the historical leap from man-control to system-control.

5 Conclusive remarks

L3 is the technology watershed in the process of enterprise achieving driverless driving. L3 level is a very critical step in achieving the leap. Because the machine has begun to replace people to make some decisions at this point.

From the perspective of technical realization of autonomous vehicles, L3 and L4 are not simple progressive relationships. This is not only a qualitative leap in technology, but also a quantitative leap in workload. The development of L3 can indeed support L4 in some aspects, including the accumulation of common technologies such as sensors and algorithms. The three technical difficulties that span from L3 to L4 are decision making, intuition, and testing and verification. The main difficulty in decision making is that getting full and complete big data is far more difficult than imagined. But, the decision systems of L4 based on deep learning are highly dependent on well-completed big data. And, the difficulty in intuition is that it is very difficult for autonomous vehicles to understand other human ' s intention and predict their behaviour. In addition, the reliability of autonomous vehicles is difficult to verify through real-world road testing and virtual world simulations.

From the perspective of the commercialization of autonomous driving, not only does it take time to do technical iterations and cost reductions, but there are also issues of consumer trust and acceptance. If the L3 product can effectively liberate the driver in certain scenarios for daily use, there is a good market prospect. Enterprises should comprehensively analyze and systematically plan application scenarios suitable for different levels of autonomous driving products. At the same time, the current stage of car companies must consider L4 autonomous driving technology research reserve.

6 Acknowledge

This research is supported by the National Natural Science Foundation of China (U1764265). The authors would like to thank all partners within for their cooperation and valuable contribution.

References

1. X. Miao. Smart Factory and the Transformation & amp; Upgrading of Equipment Manufacturing Industry. Process Automation Instrumentation. **35(03)**, 1-6 (2014)
2. F. Zhao, Z. Liu. Strategy direction of building powerful automotive industry in China. Auto Industry Research, **(10)**, 4-7 (2014)
3. Z. Liu. Fuquan Zhao's Insights on Automotive Industry [M]. Beijing: China Machine Press, 57-63 (2017)
4. F. Zhao, Z. Liu., H. Hao, T. Shi. Characteristics, trends and opportunities in changing automotive industry. Journal of Automotive Safety and Energy, **9(3)**, 233-249 (2018)
5. X. Kuang, F. Zhao, H. Hao, Z. Liu. Intelligent connected vehicles: the industrial practices and impacts on automotive value-chains in China. Asia Pacific Business Review, **24(1)**, 1-21 (2018)
6. K. Li, Y. Dai, S. Li, M, Bian. State-of-the-art and technical trends of intelligent and connected vehicles. Journal of Automotive Safety and Energy, **8(1)**, 1-14, (2017)
7. SAE J3016(TM) Taxonomy and Definitions for Terms Related to Driving Automation Systems for On-Road Motor Vehicles [s]. SAE International,18-19 (2018)
8. S. Chen, J. Hu, Y. Shi. LTE-V: A TD-LTE based V2X solution for future vehicular network. IEEE Internet of Things Journal, **3(6)**,997-1005 (2016)
9. C. Ilas. Electronic sensing technologies for autonomous ground vehicles: A review. 2013 *8th International Symposium on Advanced Topics in Electrical Engineering (ATTE)*, IEEE, 1-6 (2013)
10. J. Varghese, R. Boone. Overview of Autonomous Vehicle Sensors and Systems. *International Conference on Operations Excellence and Service Engineering,* 10-11 (2015)
11. N. Kalra, S. Paddock. Driving to safety: How many miles of driving would it take to demonstrate autonomous vehicle reliability? Transportation Research Part A: Policy and Practice, **94**, 182-193 (2016)
12. B. Kim, Y. Kashiba, S. Dai. Testing Autonomous Vehicle Software in the Virtual Prototyping Environment. IEEE Embedded Systems Letters, **9(1)**, 5-8(2017)
13. P. Lin. Why ethics matters for autonomous cars[M]. Springer, 69-85(2016)

Performance Analysis of the Sliding Mode Control for Automated Vehicle Path Tracking at Low Adhesion Surfaces

Ilya Kulikov[1,a] and Ivan Ulchenko[1]

[1]*State Research Centre "NAMI", Moscow, Russian Federation*

Abstract. The article analyses prospects of using a type of robust controllers called relay regulators for automation of vehicle lateral motion. The operation of these regulators in so-called sliding modes is considered along with the "chattering" problem caused by deviations from the "ideal" sliding mode inevitable in actual implementations. For the analysis of vehicle motion, a mathematical model was elaborated, which calculates vehicle dynamics taking into account non-linear tire-road adhesion characteristics. In the conducted study, emphasis was put on low adhesion surfaces, which can be considered as the most difficult case for automatic lateral control of a vehicle. In order to implement automated path tracking within the model, two relay regulators were elaborated differing from one another in the order of dynamics. A comparative study of these regulators was conducted by means of simulations. The regulator that had shown best performance was then tested for robustness by means of modeling, in which maneuvers on snow, ice and a mixed surface were simulated.

1 Introduction

Development of automated driving functions for vehicles constitutes a complex task, since the controlled object is non-linear, has parameter uncertainties, and operates under arbitrary conditions. Automated maneuvering at low adhesion or mixed-adhesion surfaces can be considered as most difficult control task requiring of the automated driving functions involved to be highly adaptive. The objective of this work is to develop an adaptive (robust) controller solving this control task.

In the literature, one can find at least five controller types used (or offered to be used) in automated driving functions, namely, linear-quadratic regulators (LQR), model predictive regulators (MPC), fuzzy controllers, PID (PI, PD) regulators with gain scheduling, and relay regulators operating in sliding modes.

The first two regulator types are derived from models of controlled plants. In application to automated vehicle driving, these regulators, being accompanied by rather advanced and complex observers of vehicle dynamic parameters, show good performance even in emergency maneuvers at low adhesion surfaces [1, 2]. LQR and MPC require parameters of the controlled plant to be known (identified). The stability criterion of these regulators is observability of matrices, which contain model parameters; however, this observability may substantially diminish in complicated driving conditions such as maneuvering at mixed surfaces. The second issue is linearity of LQR and MPC. Since the vehicle is non-linear object, using these regulators for its control implies linearization of the model at each time step of the controller. The third issue is high computational burden imposed by solving of regulator equations. It is admitted that existing processing units available for vehicles are not able to perform these calculations when vehicle speed increases beyond certain values.

Using conventional PID regulators for automated driving is only possible with scheduled gains. This makes regulators non-linear allowing them to handle vehicle dynamics and encompass various driving conditions. Examples presented in the literature demonstrate a good performance of such regulators in various (yet limited) tasks [3, 4]. It is clear that "schedules" are expected to be multidimensional lookup tables being functions of variables, which characterize both vehicle parameters and driving conditions (the latter are supposed to be observable). A considerable number of simulations and field-testing are needed to populate and tune those tables. However, any internal and ambient factors lying outside the range of predefined tables will question stability and adaptiveness of the regulator.

As an alternative to conventional and optimal regulators, fuzzy controllers are making their way into vehicle active safety systems and driving automation applications. However, the fuzzy approach may be problematic when it comes to the general task of driving control rather than its particular cases such as lane keeping [5] or low-speed maneuvering [6]. The first issue is that increasing of control task complexity implies expanding sets of fuzzy rules and terms. This, in turn, entails higher efforts to develop a controller and increases computational burden that may even make one resort to using separate processing units performing fuzzy

[a] Corresponding author: i.kulikov.mami@gmail.com

operations. The second issue is that the general task of driving control makes it difficult to encompass the variety of possible driving conditions and vehicle parameters by a limited set of fuzzy rules and terms. One should also take into account that the foundation of fuzzy controllers is an expert knowledge, which becomes a decisive factor of controller's robustness. The question is if this knowledge, even when provided by qualified experts, can be adequate for entire variety of possible driving situations including emergency maneuvers, when vehicle behavior becomes strongly non-linear and difficult to handle.

Relay regulators [7, 8] are free from the described shortcomings of model-based adaptive controllers (although optimal relay regulators can also be based on models) and augmented conventional regulators. Relay regulators are non-linear "by nature", which makes them suitable for control of non-linear plants with no linearizing assumptions. In the "ideal" sliding mode (explained below), both structure and operation of relay regulators only depend on the tracking error and its derivatives and are insensitive to plant's parameter uncertainties or disturbances. The structure of a relay regulator is simple, thus, not imposing high computational burden upon processing units.

The major problem of relay regulators is so called chattering [8, 9] – high frequency, sawtooth-form oscillations induced by switching of the relay element with finite frequency, which is inevitable in an actual controller having finite time step. This is a malicious phenomenon, which causes noise and vibrations and even can lead to damaging of control mechanisms and some parts of the controlled object. In the literature, a few methods are proposed to reduce chattering, namely, the boundary layer approach [10], the twister and super-twister methods [11], using of compensating state observers [8], and non-linear adjustment of the relay gain [8]. Yet another method is use of higher order sliding modes [11]. It is considered a perspective one; however, it makes regulator dynamics more complex, since it implies high orders of derivatives and integrals involved. This may entail diminishing of regulator accuracy and stability. On the other hand, these (possible) adverse effects depend on specifics of the given plant and its dynamics and only can be estimated through appropriate studying.

Despite their peculiarities and difficulties stemming from the chattering problem, sliding mode regulators are recognized as perspective robust controllers. A number of works in the field of vehicle dynamics and automated driving are dedicated to these regulators [12, 13]. A preliminary analysis suggests that the higher order sliding modes can be suitable for vehicle automated driving on the condition of providing both chattering suppression and stability of operation. This work describes an investigation of this problem putting emphasis on maneuvers performed at different low adhesion surfaces implying automated driving functions to be adaptive to ambient conditions.

A mathematical model of vehicle dynamics was elaborated in order to conduct said study. Adequacy and accuracy of the model were estimated through a comparison with experimental data obtained from tests, which were performed at actual low adhesion surfaces. An automatic path-tracking function was implemented within the model. The sequel of the article describes these models and the results of the conducted study.

The paper is organized in three main sections. Section 2 describes the developed model of vehicle dynamics. In section 3, elaboration and model analysis of the path-tracking controller are presented including a comparison of different designs of sliding mode regulators. The final section offers concluding remarks and outlines the future work.

2 Model of vehicle dynamics

A detailed description of the vehicle dynamics model used in this study can be found in [14], while in this article, it is only presented briefly. The model constitutes a system of six lumped masses, namely, the vehicle mass whose motion, projected onto the road, is considered planar, the sprung mass performing angular motion (i.e. body roll) in the lateral plane of the vehicle, and four rotating masses of the wheels. The wheel normal forces are calculated from the static equilibrium of the vehicle in both longitudinal and transversal planes.

The following coordinate systems are introduced within the model: the road-bound system, the vehicle-bound system (the variables belonging to it are denoted with indices containing "v"), and the wheel-bound systems (corresponding variables are provided with indices containing "w"). Variables related to individual wheels are also provided with indices ij, where i stands for the axis number, and j denotes number of the wheel associated with i-th axis. The model of vehicle dynamics has the form of the following equation system:

$$\begin{cases} m \cdot \left(\dot{V}_{CoG,x-v} - \omega_z \cdot V_{CoG,y-v}\right) = \sum_{i=1}^{2}\sum_{j=1}^{2} R_{x-v.ij} - F_{w,x} \\ m \cdot \left(\dot{V}_{CoG,y-v} + \omega_z \cdot V_{CoG,x-v}\right) = \sum_{i=1}^{2}\sum_{j=1}^{2} R_{y-v.ij} \\ \mathcal{J}_z \cdot \ddot{\psi}_z = \mathcal{J}_z \cdot \dot{\omega}_z = \sum_{i=1}^{2}\sum_{j=1}^{2} M_{z-v.ij} \\ \mathcal{J}_x \cdot \ddot{\varphi} = \sum_{i=1}^{2}\sum_{j=1}^{2} \left(R_{y-v.ij} \cdot h_{CoG}\right) - \varphi \sum_{i=1}^{2} c_{\varphi i} - \dot{\varphi} \sum_{i=1}^{2} d_{\varphi i} \end{cases} \quad (1)$$

where m is the vehicle mass, $V_{CoG,x-v}$ and $V_{CoG,y-v}$ are the components of the vehicle velocity vector in the center of gravity (CoG) projected onto vehicle's longitudinal and lateral axes respectively, ω_z and ψ_z are the vehicle yaw rate and yaw angle respectively, $R_{x-v.ij}$ and $R_{y-v.ij}$ are the components of the tire force projected onto vehicle's longitudinal and lateral axes respectively, $M_{z-v.ij}$ is the yawing moment exerted by the wheel, $\mathcal{J}_z$ is the vehicle yaw inertia, $\mathcal{J}_x$ is the roll inertia of the sprung mass, φ is the body roll angle, h_{CoG} is the CoG height, $c_{\varphi i}$ and $d_{\varphi i}$ are the roll stiffness and damping of the suspension.

In the simulations conducted in this work, the vehicle was maneuvering with no combined tire slip (i.e. slip, which consists of longitudinal and side components both having substantial magnitudes). In that case, the tire forces in the wheel-bound coordinate system and the tire aligning moment (M_{z-w}) can be calculated (with some simplifying assumptions) by the following functions: $R_{x-w} = R_z \cdot \mu_x(S_x)$, $R_{y-w} = R_z \cdot \mu_y(\alpha)$, $M_{z-w} = R_{y-w} \cdot t_0(\alpha)$. In these formulae, μ_x and μ_y are longitudinal and lateral tire-road adhesion coefficients respectively, S_x is the longitudinal tire slip, α is the sideslip angle, and t_0 is the tire pneumatic trail.

The components of the tire slip are calculated by expressions, which include linear and angular velocities calculated by the equation system (1). These expressions are omitted here and can be found in [14]. For calculation of the adhesion coefficients and the pneumatic trail, an empirical tire model called the Magic Formula [15] was employed.

The model was verified for a passenger vehicle [14] (the gross mass 2550 kg), which is studied in this work.

3 Elaboration and model analysis of the path-tracking control

The path-tracking regulator operating principle is to minimize deviation between the vehicle driving direction and the reference traveling path. Figure 1 shows a schematic used for calculation of this deviation. In order to make the regulator respond to changes in the required path in advance, the deviation is observed at the point lying ahead of the vehicle, on the extension of its longitudinal axis. The distance between the vehicle CoG and the observation point is denoted "x-obs". The required vehicle trajectory is defined by an array of coordinates, which is approximated by the linear piecewise method. Thus, each fragment of the trajectory is presented by a linear algebraic equation. A similar equation describes the line being normal to the vehicle longitudinal axis and intersecting it in the observation point. Combining equation of this orthogonal line with the equation of the trajectory's piece, which is nearest to the vehicle, allows obtaining the intersection point of these two lines and, consequently, yields the deviation Δy. Calculation of this deviation is based on rather simple geometry; however, the corresponding equations are bulky and therefore omitted here.

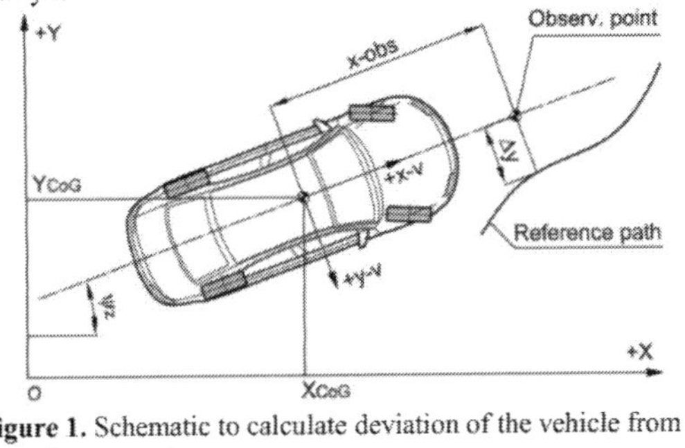

Figure 1. Schematic to calculate deviation of the vehicle from the reference path.

Implementation of a control system based on a relay regulator arises the question how to translate a relay signal into the command for an actuator. In the case of vehicle trajectory control, the latter constitutes a steering mechanism. For its automatic control, the steering angle command is required. It is obvious that this command should be a continuous signal. Moreover, neither steering speed nor angular acceleration can be discontinuous signals. This implies placing series of integrators between the relay signal and the steering angle command. However, increasing order of integration may entail degrading of regulator stability and tracking quality. Thus the number of integrators should be a trade-off between smoothness of the regulator output signal and the stability (and quality) of its operation. During the study, it was established that the minimum admissible order of integration is second, which implies relay switching to be at the level of angular acceleration signal. The mathematical description of the resulted regulator (hereafter called the Regulator-1) reads as follows:

$$\ddot{\delta}_{SW} = \gamma \cdot sgn(s).$$

where δ_{SW} is the steering wheel angle. Instead of sgn function, a relay element may be employed. The integrators placed after the relay element have limits, which define the maximum steering speed (in this study, the permissible speed range was ±300 °/s) and the maximum steering angle (was constrained by the range ±200°).

In theory, the relay gain γ should be infinite that would have resulted in infinite switching frequency yielding the "ideal" sliding mode. It is obvious that this cannot be implemented in actual applications; therefore, the value of gain is set finite yet sufficiently large. Studying the regulators described in this work allowed to conclude that increasing the gain up to 1e5 improves the regulator performance. Above that value, no changes were observed.

$s = f(\Delta y)$ is a function defining so-called hyperplane of the sliding mode or the sliding surface. In the "ideal" sliding mode, the phase trajectory of the controlled plant follows along this surface. In actual sliding modes, phase trajectories dwell in some vicinity of the sliding surface. For the Regulator-1, this surface was defined as follows:

$$s = c_1\Delta y + c_2\Delta\dot{y} + c_3\Delta\ddot{y}$$

where c_1, c_2, c_3 are tunable parameters defining the shape of the sliding surface.

For model analysis of the path-tracking regulators, different maneuvers were conducted, one of which implied driving along the test track shown in the uppermost graph in Figure 2. The three bends of the track have the following radii (from left to right): 35 m, 40 m, and 40 m. The vehicle was driving at a constant speed, which, at a given surface, required rather intensive steering in order to follow the trajectory with an acceptable accuracy. On the other hand, the speed was not high enough to bring the vehicle into a skid or drift, because this would have required special measures to maintain the vehicle yaw stability, which was beyond the scope of the study.

Figure 2 shows the modeling results with involvement of the Regulator-1. The graphs show the following

https://doi.org/10.1051/matecconf/201929601003

variables (from top to bottom): vehicle trajectory, the output signal of the relay element, the steering angle signal. In these graphs, one can see that the regulator provides satisfactory accuracy of path tracking; however, the steering angle signal contains a pronounced chattering, which is undesirable for actual implementation of the control system within a vehicle.

For elimination of the chattering, the regulator order (i.e. the order of both input derivatives and output integrators) was increased by one. The resulted design is referred to as the Regulator-2. Higher orders of the derivatives and integrators entail questions of the noises at the regulator input and stability of its operation. The first question requires an additional investigation, which lies beyond the scope of the described study. The stability issue was investigated by simulations of the above maneuver. Figure 3 shows the simulation results for the same driving conditions as were used previously for the Regulator-1.

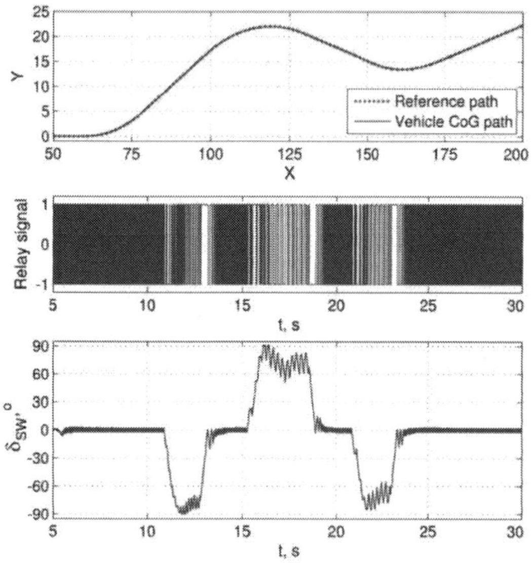

Figure 2. Simulation results for the test track covered with snow. Regulator-1. Vehicle speed 35 km/h.

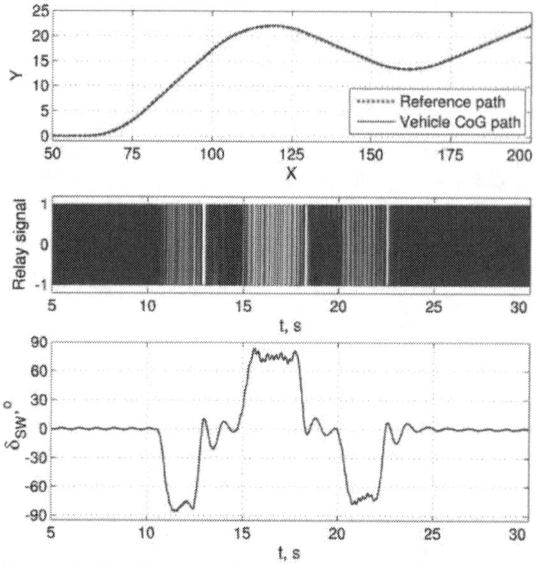

Figure 3. Simulation results for the test track covered with snow. Regulator-2. Vehicle speed 35 km/h.

From these graphs, one can make the following observations. Using of the Regulator-2 eliminates chattering from the output signal. In this aspect, the Regulator-2 makes a possible candidate for implementation within an actual system. Both the regulator operation and vehicle motion are stable. The path tracking accuracy degraded negligibly. Oscillatory behavior of steering transients has increased, which was expected due to higher integration order (i.e. increased phase shifting). However, the oscillations have relatively small amplitude and fade rapidly. Self-oscillations of smaller amplitude and higher frequency are observed when the mean value of steering angle is close to constant (see also Figure 4). Although this is not chattering, such oscillations of the steering wheel are undesirable. Therefore, in the sequel of the research, methods to damp these oscillations are to be found.

In order to investigate robustness of the sliding mode, i.e. regulator's behavior in differing and complicated driving conditions, the Regulator-2 was tested at ice-covered track and at mixed surface imitating a snowy road with icy patches. The simulation results for both surface types are shown in Figures 4 and 5. In the trajectory graph for the mixed surface, the grey areas depict mentioned icy patches.

In the both cases, the regulator has provided stable operation and good tracking accuracy. At the mixed surface, deviations from the reference path are slightly higher than at the homogeneous snowy surface. At the icy surface, one can notice above-mentioned self-oscillations of the steering angle signal, while at the mixed surface there are no such oscillations due to absence of steady state steering. Note that in all the described simulations, the regulator parameters (i.e. gains) remained unchanged.

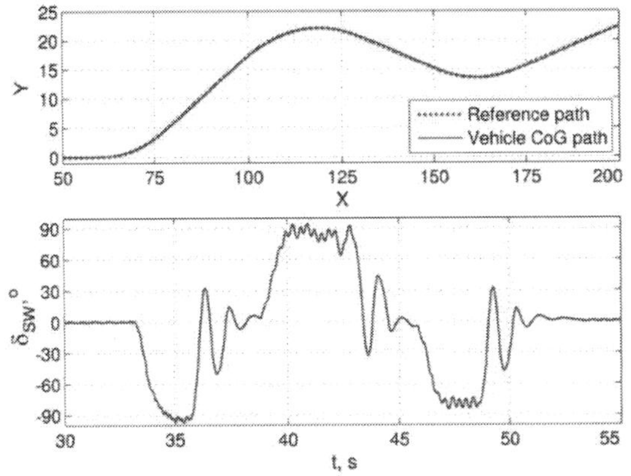

Figure 4. Simulation results for the test track covered with ice. Regulator-2. Vehicle speed 28 km/h.

Table 1 summarizes path-tracking performance of the elaborated regulators in the above maneuvers at different surfaces. Two parameters are used to characterize the tracking accuracy, namely, the mean square value (MSV, in meters) and the maximum value (Max., in meters) for the deviation of the CoG trajectory from the reference path (excluding the first straight-line part of the test track where no steering is required).

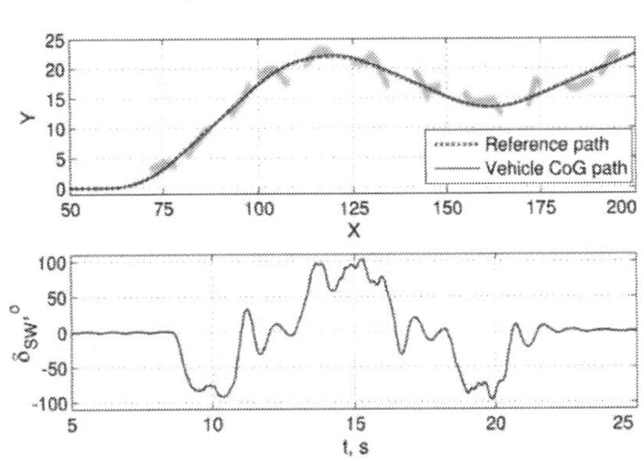

Figure 5. Simulation results for the test track with mixed surface (snow and ice). Regulator-2. Vehicle speed 35 km/h.

Table 1. Accuracy parameters of the path-tracking regulators at different surfaces.

Reg.#	Snow		Ice		Mixed	
	MSV	Max.	MSV	Max.	MSV	Max.
1	0.05	< 0.1	n/a	n/a	n/a	n/a
2	0.065	< 0.15	0.13	< 0.28	0.1	< 0.2

4 Conclusions and future work

The conducted study has shown that the sliding mode control, even non-ideal, provides a considerable robustness. A relay regulator, having constant parameters and no underlying plant model, performs intensive vehicle maneuvering at snow, ice and mixed surface with good path tracking accuracy. The relay regulator with second order integration of the output signal operates with chattering, which most likely would be unacceptable in actual implementations. Increasing of the regulator order (both input derivative and output integration) by one allows to eliminate chattering. Simulation results show that despite high integration order the regulator operation is stable, while the tracking accuracy degrades negligibly. However, increasing of the regulator's order questions the calculation quality of derivatives, since some noise is inevitable in feedback signals. In this regard, the usually claimed robustness of sliding modes against disturbances becomes a crucial feature to be the subject of further research.

Acknowledgments

The article was prepared under the agreement #14.625.21.0043 with the Ministry of Education and Science of the Russian Federation (unique project identifier RFMEF162517X0043).

References

1. M. Alirezaei, S.T.H. Jansen, A.J.C. Schmeitz, A.K Madhusudhanan, Collision Avoidance System using State Dependent Riccati Equation Technique: An Experimental Robustness Evaluation, *13th Int. Symposium on Advanced Vehicle Control (AVEC'16)*, pp. 127-132 (2016).
2. T. Keviczky, P. Falcone, F. Borrelli, J. Asgari, D. Hrovat, Predictive control approach to autonomous vehicle steering, *American Control Conference*, pp. 6-15 (2006), doi: 10.1109/ACC.2006.1657458.
3. M. Keller, C. Haß, A. Seewald, T. Bertram, A vehicle lateral control approach for collision avoidance by emergency steering maneuvers, *6th Int. Munich Chassis Symposium*, pp. 175-197 (2015), doi: 10.1007/978-3-658-09711-0_15.
4. P. Zhao, J. Chen, Y. Song, X. Tao, T. Xu, T. Mei, Design of a Control System for an Autonomous Vehicle Based on Adaptive-PID, Int. Journal of Advanced Robotic Systems **9**(2), 1 (2012), doi: 10.5772/51314
5. N.C. Basjaruddin, Kuspriyanto, Suhendar, D. Saefudin, S.A. Aryani, Lane Keeping Assist System Based on Fuzzy Logic, *Int. Electronics Symposium (IES)*, pp. 116-119 (2015), doi: 10.1109/ELECSYM.2015.7380824
6. J. Pérez, V. Milanés, E. Onieva, Cascade Architecture for Lateral Control in Autonomous Vehicles, IEEE Transactions on Intelligent Transportation Systems, **12**(1), pp. 73-82 (2011), doi: 10.1109/TITS.2010.2060722.
7. S.V. Emelyanov, *Automatic Control Systems with Variable Structure* (in Russian), Moscow, "Nauka" (1967).
8. V. Utkin, J. Guldner, J. Shi, *Sliding Mode Control in Electromechanical Systems*, Second Edition, Taylor & Francis Group, LLC (2009).
9. A. Levant, Chattering Analysis, *IEEE Transactions on Automatic Control*, **55**(6), pp. 1380-1389 (2010), doi: 10.1109/TAC.2010.2041973.
10. J. J. Slotine, S. S. Sastry, Tracking control of non-linear systems using sliding surfaces, with application to robot manipulators, International Journal of Control, **38**(2), pp. 465–492 (1983), doi:10.1080/00207178308933088.

11. L. Fridman, A. Levant, Higher-Order Sliding Modes, book chapter (*Sliding Mode Control in Engineering*, chapter 3), Inc. Marcel Dekker (2002).
12. A. Ferrara, *Sliding mode control of vehicle dynamics*, London: Institution of Engineering and Technology (2017), doi:10.1049/PBTR005E.
13. G. Tagne, R. Talj, A. Charara, Higher-Order Sliding Mode Control for Lateral Dynamics of Autonomous Vehicles, with Experimental Validation, *IEEE Intelligent Vehicles Symposium (IV)* (2013), doi: 10.1109/IVS.2013.6629545.
14. I.A. Kulikov, S.V. Bakhmutov, A.A. Barashkov, An Investigation of Vehicle Dynamics Concerning Active Safety Systems by Simulations and Driving Tests, *Proceedings of NAMI*, **265**, pp. 53-65 (2016).
15. H.B. Pacejka, I. Besselink, *Tire and vehicle dynamics*, Third Edition, Elsevier Ltd., pp. 165-183 (2012).

Prediction of Ship Traffic Volume in Jiujiang Port Based on Genetic Wavelet Neural Network

Shouwei XIE [1,a] and Yadong YANG[2]

[1]*College of Wuhan University of Technology, Hubei Provincial Key Laboratory of Inland navigation technique, china*
[2]*College of Wuhan University of Technology, Hubei Provincial Key Laboratory of Inland navigation technique, china*

Abstract. In recent years, the traffic volume of the Yangtze River has increased dramatically. In order to provide more favorable assistance to port planning and traffic management, the accuracy of port ship traffic volume prediction is very important. In this paper, genetic algorithm and wavelet analysis and neural network are used to construct the genetic wavelet neural network model prediction model, and BP neural network prediction model is established. The ship volume of Jiujiang Port is used as experimental data to simulate and analyze. The results show that the prediction accuracy of the genetic wavelet neural network prediction model is significantly higher than that of the BP neural network prediction model. It is proved that the genetic wavelet neural network has broad application prospects for ship traffic flow prediction in the Yangtze River port. This method has practical application significance.

1 The introduction

These In recent years, the domestic economic development trend has become better and better, and various industries have also developed rapidly. Among them, the development of waterways has become more and more prominent. In order to meet the needs of economic growth, the traffic volume of the Yangtze River has increased sharply in recent years, and the development of China's waterway transportation is prominent at this stage. Contradiction is the contradiction between advanced navigation technology and backward port transportation management. In order to solve this contradiction, it is necessary to strengthen port ship management and preparation[1]. Accurately predicting the traffic volume of port ships can alleviate the pressure of shipping traffic and the maritime department and port transportation management departments. It can also formulate more detailed berthing, berthing and piloting plans to fundamentally solve the problem of congestion in port transportation and improve the application of ships. effectiveness.

There are two general aspects to the prediction of ship traffic volume:

(1) Prediction of time series;

(2) Predictions in terms of impact factors;

The specific methods include regression prediction, time series prediction, and gray theory prediction. However, these methods have some limitations in predicting the ship traffic volume, and the prediction results have large errors[2]. The neural network prediction model has a simple structure, the application method is relatively simple, and the neural network information synthesis ability is strong and the fault tolerance is high. The prediction results are relatively accurate. Wavelet neural network combines the advantages of wavelet analysis and neural network in dealing with nonlinear problems, and solves the blindness of BP neural network structure design and the problem of easy to fall into local optimum[3].Therefore, this paper establishes a genetic wavelet neural network prediction model and a BP neural network prediction model to predict the traffic volume of Jiujiang Port.

2 Establishment of prediction model

Prediction compared with the simple BP neural network Fan Qingbo used BP neural network and Markov model, constructed BP neural network-Markov prediction model, and optimized the model by particle swarm optimization. Finally, the monthly standard ship traffic flow of Wuhan Yangtze River Bridge was taken as a sample[4]. model, the conclusion that the BP neural network-Markov prediction model has higher accuracy is obtained. In addition, some scholars use wavelet neural network [5] and other methods. This paper will use genetic algorithm and BP. The neural network and wavelet neural network are combined to construct the ship traffic flow prediction model, and it is applied to the ship traffic flow in Jiujiang Port. Then the results are compared, and the difference between the two methods in the prediction results and the method are discussed. Specific application.

2.1 Establishment of genetic wavelet neural network prediction model

[a] Corresponding author: 1036849174@qq.com

 https://doi.org/10.1051/matecconf/201929601004

2.1.1 Wavelet neural network

Wavelet neural network is a breakthrough by wavelet analysis in recent years, research on the basis of a kind of artificial neural network is put forward, which is based on the theory of wavelet analysis and wavelet transform is a kind of hierarchical structure made by new type of multi-resolution and artificial neural network model, which USES nonlinear wavelet base instead of the usual nonlinear sigmoid function, the signal is described as table by the selection of wavelet base linear superposition, the corresponding input layer to hidden layer weights and threshold of hidden layer, respectively, by the expansion of the scale of the wavelet function is replaced by the translation factor and time factor. Wavelet function has good localization property and has penetrated into various fields. Wavelet transform is a time-frequency localization analysis method whose window size is fixed but whose shape, time window and frequency window can be changedm[6-7]. used because of its localized characteristics. Its basic principle is a layered, multi-resolution new artificial neural network model composed of wavelet analysis and wavelet transform. Its model structure is shown in figure 1.

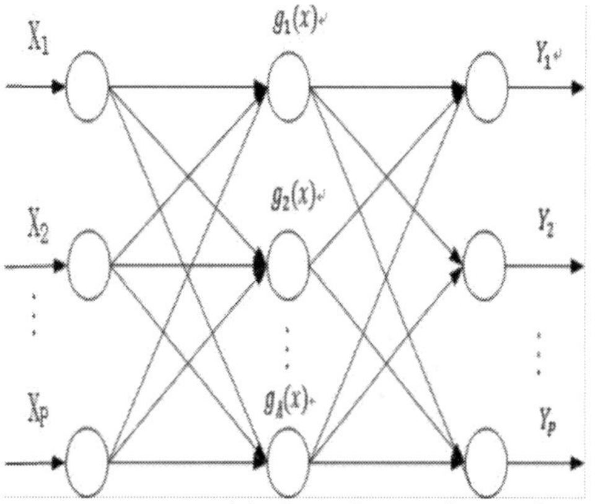

Figure 1. structure model of wavelet neural network

Wavelet neural network is a three-layer function, which is the input layer, hidden layer and output layer. Figure X shows the input sample of the input layer, g(x) is the wavelet function. It includes the scale scaling factor, the time shift factor, and y is the output layer. The number of hidden layer units of the wavelet neural network can be adaptively determined as follows: First, take the wavelet network neural network hidden layer unit number M is 1, after learning iteration several times, if it meets the error condition, stop iteration. If the maximum learning learning time is reached, the error condition is still not satisfied, then M + 1; repeat the above steps Until the error condition is met, this can overcome the blindness of the traditional neural network structure design.

2.1.2 BP neural network

BP neural network is also called error back propagation network, which has the advantages of strong learning ability, good nonlinear mapping ability and good fault tolerance. During the running of the algorithm, the information goes forward and the error propagates back to correct the network [8]. The algorithm core of the network is a step method (the steepest descent method). By optimizing the connection weight between layers, the sum of squared errors between the actual output value of the neural network and the ideal output is minimized.

2.1.3 Genetic algorithm (ga)

Genetic algorithm is a global optimization search algorithm that simulates the biological evolution process of genetic selection and natural elimination [9].The operation of this algorithm has selection, crossover, mutation, and three operations. The genetic algorithm is simple to operate, and its parallel search ability is integrated into the nerve. In the network weight training of the network, it can search for the initial point with global ergodicity to ensure network training convergence and reduce training time. When genetic algorithm is applied to the optimization of network weights, its implicit parallel performance can overcome the difficulty that weights are easily trapped in local minima in the process of training, and it has the powerful function mapping and approximation ability of neural networks [10].

2.2 Construction of prediction model

2.2.1 Construction of genetic wavelet neural network model

Processing of input data
Based on common sense and basic statistical knowledge, the raw data is basically processed, the erroneous data is eliminated, and then each input data has different units and value dimensions, so each input data needs to be normalized.
Chromosome coding
The individual coding method is real number coding, and each individual is a real number string, which consists of an input layer and an intermediate layer link weight, an intermediate layer scale factor, a b, and an intermediate layer and an output layer connection weight. The individual contains all the weights and thresholds of the neural network. The total weights and thresholds in this paper are $5 \times 5 + 5 + 5 + 5 \times 1 = 40$, which is 40-bit code.
Wavelet function selection
For the selection of wavelet functions in wavelet networks, there is no unified theory or method to determine which type of wavelet function or scaling function is suitable as the excitation function of neurons in different networks, usually based on experience and actual situation. Of course, you can also learn from some of the experiences in wavelet analysis. In this paper, Morlet wavelet function is used as the excitation function of the network hidden layer by analyzing the literature and collecting data.
Fitness function

 https://doi.org/10.1051/matecconf/201929601004

According to the initial weight and threshold of the wavelet neural network, the training data is used to train the wavelet neural network and predict the output of the system. The sum of the absolute values of the errors between the predicted output and the expected output is taken as the individual fitness value F. The calculation formula is :

$$F = \sum_{i=1}^{n} abs(y_i - O_i) = \pi r^2 \qquad (1)$$

Where, n is the number of network output nodes; y_i is the expected output of the i-th node of BP neural network, and O_i is the predicted output of the i-th node. The volume chart of ship traffic volume prediction algorithm is shown in figure 2.

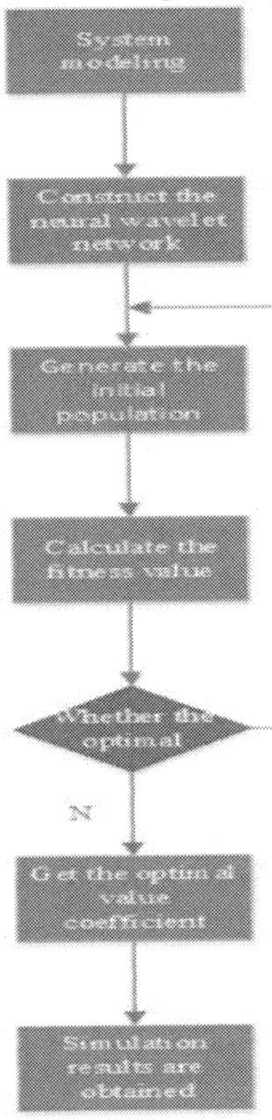

Figure 2 . genetic wavelet neural prediction model

2.2.2 Construction of BP neural network model optimized by genetic algorithm

Genetic algorithm optimization BP neural network is divided into three parts: BP neural network structure determination, genetic algorithm optimization and BP neural network prediction. The structure determining part of the BP neural network determines the BP neural network structure according to the number of input and output parameters of the fitting function, and then determines the length of the genetic algorithm individual. Genetic algorithm optimization uses genetic algorithm to optimize the weight and threshold of BP neural network. Each individual in the population contains all the weights and thresholds of a network. The individual calculates the fitness value of the individual through the fitness function. The genetic algorithm selects , crossover and mutation operations find the optimal fitness value corresponding to the individual. BP neural network prediction uses genetic algorithm to obtain the optimal individual's initial weight and threshold assignment to the network, and the network predicts the function output after training. The BP neural network model optimized by genetic algorithm is shown in Figure 3.

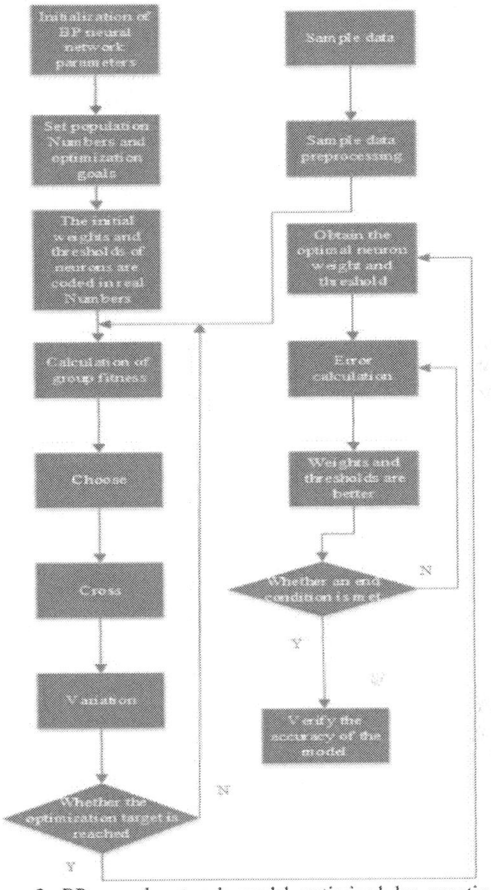

Figure 3. BP neural network model optimized by genetic algorithm

2.3 Data sampling and processing

Considering the many factors affecting the traffic volume of port ships, this paper uses the experimental data samples from January 2015 to January 2017 to predict the traffic volume of ships in Jiujiang Port in January 2018. In order to better characterize the concept of ship traffic volume, the port ship traffic volume has a more intuitive performance, and then accurately reflect the scale of the ship traffic volume in the waterway and the importance of the waterway, and convert the original data ship traffic volume into The weighted traffic volume of the ship, known as the standard ship, is also directly expressed as a standard ship. The conversion method is as shown in

Table 1[11]. by the standard ship converted jiujiang port traffic volume are shown in table 2.

Table 1 standard ship conversion coefficient table.

Gross tonnage of ship	The captain	Conversion factor
<100	<30	0.25
100~499	30~50	0.5
500~2999	50~90	1
3000~5999	90~115	1.18
6000~9999	115~135	1.41
10000~14999	135~155	1.7
15000~19999	155~170	2.0
20000~29999	170~190	2.25
30000~39999	195~215	2.5
40000~59999	215~246	3.0
>60000	>246	4.0

Table 2 ship traffic volume of jiujiang port from January 2015 to June 2017

Time	The sorting	Vessel traffic volume (standard vessel)	Time	The sorting	Vessel traffic volume (standard vessel)
2015.1	1	910	2016.4	16	1030
2015.2	2	644	2016.5	17	919
2015.3	3	918	2016.6	18	889
2015.4	4	939	2016.7	19	940
2015.5	5	922	2016.8	20	820
2015.6	6	1037	2016.9	21	789
2015.7	7	1028	2016.10	22	1050
2015.8	8	1039	2016.11	23	1039
2015.9	9	1015	2016.12	24	1080
2015.10	10	1190	2017.1	25	1190
2015.11	11	1020	2017.2	26	949
2015.12	12	449	2017.3	27	920
2016.1	13	1060	2017.4	28	1110
2016.2	14	689	2017.5	29	1219
2016.3	15	1070	2017.6	30	1009

3 Simulation experiment and result research

The simulation experiment was carried out in the MATLAB environment. The 24 sets of data in the 30 sets of port ship traffic volume data were network trained, and the remaining 6 groups were used as test data. The obtained neural network is trained by using the obtained training data to construct a neural network prediction model. The traffic volume data of the first six months of Jiujiang Port is selected to predict the traffic volume for the next month. The number of input nodes is 6, and the output node is 1. When the initial parameters of the neural network are optimized by GA, the maximum number of iterations is set to 100, and the population size is 100. The learning rates lr1 and lr2 are 0.01 and 0.001, respectively. Then the results of the training were tested on the traffic volume of Jiujiang Port from January to June 2018 .

By using the model established by genetic neural network, the weights and correlation coefficients optimized by the genetic algorithm are assigned to the wavelet neural network. The absolute error percentage of the predicted and actual values of the test samples is used to evaluate the network prediction effect. Figure 4 is a graph of the absolute percentage of error for the network predictive output and expected output. It can be seen from Figure 4 that the prediction error values of the genetic wavelet neural network are distributed between -0.08 and 0.06.

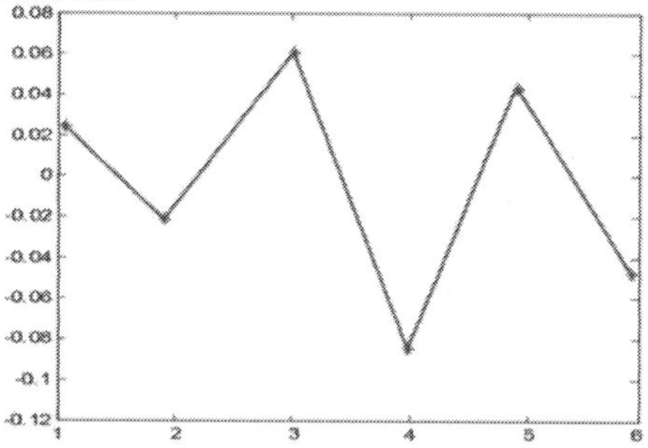

Figure 4. error percentage of prediction results of genetic wavelet neural network

Figure 5 below is the error result obtained by using Matlab operation of BP neural network optimized by genetic network. According to figure 5, the prediction error of genetic BP neural network are distributed between -0.15 and 0.3.

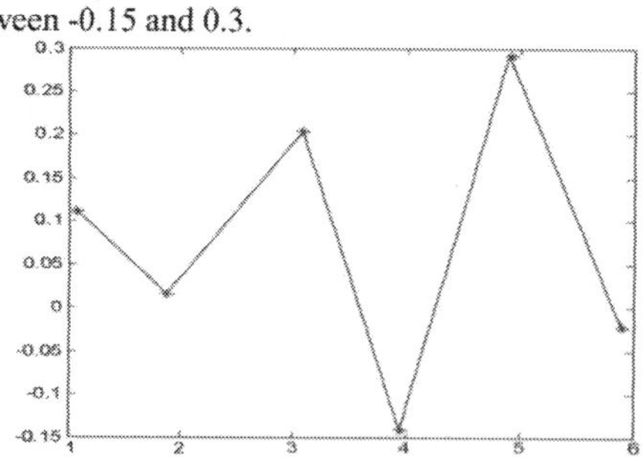

Figure 5. BP neural network result error percentage optimized by genetic algorithm.

It can be seen from Fig. 4 and Fig. 4 that the prediction error of genetic wavelet neural network is significantly smaller than that of BP neural network optimized by genetic algorithm. Using the trained genetic wavelet neural network and genetic algorithm optimized BP neural network to predict the ship traffic volume of Jiujiang Port from January to June 2018, the anti-normalization of the prediction results, and the predicted ship traffic volume shown in Table 3.

Table 3 forecasts ship traffic volume

Time	Actual traffic volume	Wavelet genetic neural predicti-on	Predic-tion error /%	BP genetic neural network prediction	Predic-tion error /%
2018-01	842	824	2.1	934	11
2018-02	481	490	-2	476	1
2018-03	833	783	6	666	20
2018-04	782	845	-8	899	-15
2018-05	783	814	4	548	30
2018-06	799	832	-4.2	823	-3

4 Conclusion

The analysis results of the above experiments show that the genetic wavelet neural network model is more accurate in predicting the traffic volume of the port than the genetic BP neural network prediction model, and the volatility in the prediction process is also small, which indicates the wavelet transform ratio. The traditional activation function in BP neural network is easier to obtain accurate prediction results. The wavelet neural network optimized by genetic algorithm tends to be globally optimal, and can improve the accuracy of prediction results.

In this paper, the characteristics of genetic algorithm are used to optimize the wavelet neural network, and its global fast optimization characteristics are combined with the characteristics of time-frequency locality of wavelet analysis to establish the genetic wavelet formed by genetic algorithm wavelet analysis and neural network. Neural network prediction models to predict port traffic volume. The experimental results show that the model has both the global fast optimization ability of genetic algorithm and the good time-frequency locality of wavelet transform, which can improve the prediction accuracy and propose an effective solution to solve the randomization problem of ship traffic volume

References

1. Zhang Renchu. Research on traffic volume prediction of ships inningbo port [D]. Dalian: dalian maritime university, 2008.
2. Li Hongxi, Fu Yuhui, Zhang Renchu. Prediction of port ship traffic volume [J]. Journal of dalian maritime university, 2009,35 (3) : 40-42.
3. HUANG Xiaohui, ZHANG Cuifang. Short-term traffic flow prediction based on cuckoo algorithm optimized wavelet neural network[J]. Computer Applications and Software, 2017, 34 (3): 238-242.
4. Fan Qingbo, Jiang Fucai, Ma Quanfang et al. BP neural network based on PSO-Markov ship traffic flow prediction model[J].Journal of Shanghai Maritime University, 2018,39(2):22-54.
5. [8] Meng Fanlin. Short-term prediction of ship traffic flow in VTS waters based on wavelet neural network[J]. China Maritime, 2017, (4): 45-47
6. Yu Xiaojiang, Hu Xuejun, Hu Yujin, et al. Intelligent adaptive control of vehicle spacing based on fuzzy neural network [J]. Journal of huazhong university of science and technology: natural science 2007, 35(9) : 22-24.
7. BARA A， DALE S， RUSU C， et al． DC Electrical Drive Control with Fuzzy System ［C］ Processing of the 13th International Conference on Engineering of Modern Electric System， ICEMES 2015， Oradea， Romania， 2015:207 －201.
8. Wang Chaoxue, Kong Yueping, Dong Lili, et al. Intelligent Optimization Algorithm and Application [M]. Xi'an: North University Press, 2012.
9. Lu Feng, Xu Jianhua, Wang Zhanyong. Hefei City water demand prediction genetic wavelet nerve Network model research [J]. Surveying and Mapping Science, 2013, 38(5): 28-31.
10. Wang Caoxue , Kong Yue ping, etal. Intelligent optimizaton algorithm and its application [M]. Xi 'an: northwestern university press, 2012.
11. Wu haolin, Zhu Jun. Maritime traffic engineering [M]. Dalian: dalian maritime university press, 2004.

Factors Affecting Crash Frequencies: A Negative Binomial Regression Based Analysis of Indus Highway, Pakistan

Rafi Ullah Khan[1], Jingbo Yin[1,a] and Faluk Shair Mustafa[1]
1 School of Naval, Architecture, Ocean and Civil Engineering, Shanghai Jiao Tong University, Shanghai 200240, China

Abstract. The increase in vehicular traffic have also increased the highway crash frequency with the passage of time. Improvements in highway safety is of vital importance as it could save vast life and monetary losses. The highway crash frequency analysis of major Pakistani highways is a subject less discovered and many important strategic and trade routes are not studied in this regard. This study is aimed to analyze the crash frequency and the prominent factors that cause these crashes on a 302 km section of Indus highway; one of the most important trade routes of the country. Eight years' data from 2011 till 2018 was arranged into 19 variables where the crash frequency is set as dependent variable, while the eighteen prominent causation factors as independent variables. The tool used for analysis was negative binomial regression being run in the SPSS software. The results indicate that the driver's behavior, understanding & risk recognition, negligence and law adherence have a significant effect on the crash frequency. Furthermore, highway crash frequency significantly increases with increase in highway segment lengths, number of lanes and lane widths. Similarly, the highway crash frequency significantly enhances when the light, pavement surface and climate condition gets deteriorated. The results of this study are of vital importance to government, transportation companies and general public in order to recognize the most important accident causing factors and devise the transport policies, rules and behaviors accordingly.

1 Introduction

The severity level of crashes are subject to the cumulative influences of various observed and unobserved factors [1]. These factors could either be engineering or non-engineering. The engineering factors includes geometric and structural design of the roadway, traffic operation & management and the pavement surface condition [2]. The non-engineering factors stands for the behavioral patterns observed among drivers, various other human factors, effects of environment, seasonal and diurnal variations.

An overview of the past literature shows that in analysis of the highway crashes, the most abundantly used variables are related to the engineering and design. While the effect of factors other than engineering, specifically that of driver behavior have been significantly ignored [3]. Even though the behavior oriented studies reflect that driver behavior is the most prominent contributory factor, it is still not properly incorporated into crash risk modelling [4]. The reason for this is the unavailability of a profound and reliable data collection method and source that could collect the data related to various trends in the driver's behavior [5]. The occurrence of a crash could be attributed to the driver being under the effect of alcohol or drugs, fatigue and carelessness.

The lack of data in this regard could be attributed to the reason that it is not the responsibility of federal or state highway agencies to take into record the behavioral patterns of the drivers involved in each accident. The data on the speed, lanes, distractions, disparities, and risky behaviors of drivers could be well found in the police reports of each accident [6].

The other prominent factors that effects the intensity and frequency of highway crashes are the sudden changes in the geometries of these highways, hindrances on the road way and the prevailing climatic conditions at the site of crash [7]. The climatic or weather alterations includes the onset of rains, tempests, smog, fog, slick pavements and inadequate light situations. All these factors affect the driver's ability to avoid any crash and collision on the road [8]. Moreover, the failures associated with brakes, axels, tires and other perfunctory culpabilities can result in crashes.

[a] Corresponding author: jingboyin@sjtu.edu.cn

The tremendous increase in the motorized vehicles, the presence of a mix of motorized and non-motorized traffic and the susceptible transport modes like motorcycles & rickshaws have increased the highway crashes manifold in Pakistan [9]. The country has to face around a hundred billion rupees of annual losses in terms of fatalities, injuries and communal defies resulting from highway crashes [10]. The highway crashes result in more than thirty thousand fatalities and around half a million injuries every year. Studies suggest among the contributory crash factors, 60% are attributed to the human factor, 30% to environmental factors, while 10% are attributed to the various mechanical faults associated with vehicles [11]. A better understanding of the contributory crash factors is pivotal in formulating better safety processes and policies.

This study is one of the few attempts to relate the highway crash frequency to various contributory factors like roadway geometries, environmental aspects, time variation and various human behavioral patterns. This study will evaluate the connectivity between road crashes initiated by geometric and traffic factors like number, type and width of lanes, shoulders, medians, access points, U-turns, curves, speed limits and traffic percentages. While the other factors taken into account were the age, gender, behavior patterns of the drivers, specific time, weather, and season at the time of crash along the vehicle conditions.

2 Literature review

Better understanding of the various contributory factors to crashes can result in better policies and measures for their reduction and management. The geometry of a road segment has a significant effect on the frequency and severity of crashes occurring in that segment [12]. The per unit length assessment of traffic crashes using Tobit regression model indicates that the pavement quality plays a key role in the crash occurrence [13].

A study of the Florida highway states that among the human factors analysis, 94% of the fatal crashes are of the kind where drivers were under the effects of alcohol [14]. The analysis of highway accidents in Pakistan reveals that unskilled drivers, heavily overloaded vehicles, pavement surface conditions and the use of mobile phones while driving are among the significant contributory factors [15]. A study conducted through multiple linear regression indicates that among the road crashes in Jordan, pavement surface and lighting conditions have the least effect on crash occurrence [16].

The behavior of a driver can be elaborated by his professional & social values along proclivity towards safety and risk [17]. The way a person looks at a risk could be affected by his society and surroundings. It could be better understood by observing that some of the drivers observe and follow other drivers in the surrounding and community instead of following the highway features and rules [18]. The expectation which a community or specific group of people in the surrounding keeps also affects the way driver perceives risk [19].

The effect of a driver's behavior on a crash in the count related models could be replaced by substitute variables like that for over speeding [20]. Same technique could be adopted for driving under the influence of alcohol and drug [21]. Though alcohol and drugs are a type of societal crimes, however they are highly linked to the highway crashes and must be considered.

In order to evaluate the highway crash frequencies and the various contributing factors, various research methods and tools have been used over the period of time. In the past these techniques were simple and based on linear regressions mostly, however with the passage of time the normality of error, residuals and the other issues associated with linear relationship between various dependent and independent variables were sorted out [22]. Therefore, to accommodate the multifaceted relationship between various variables, Poisson regression models were started using which were based on the consideration of exponential relationship between the involved variables.

However, the dispersion of the data which required the equality of mean and variance for Poisson regression proved to be a limitation. This issue has been sorted out by the introduction of Negative binomial regression, which is considered independent of the over and under-dispersion limits. Though negative binomial regression model has some limitations associated with the multivariate variables, however it is considered very strong and efficient tool in evaluating the crash, fatality, and injuries related statistical data [2]. However, the use of this technique to evaluate the highway crashes data of Pakistan is very limited and various studies needs to be conducted evaluating various factors that affects the occurrence of crashes in land, air and maritime transportation.

3 Methodology & data collection

The purpose of this study was to determine those geometric, human, weather and vehicle factors which played effective role in the occurrence of crashes at the selected segment of the road. The relationship between the crash occurrence frequency and the engineering and non-engineering factors have been analyzed using the negative binomial regression. For this study, a segment of the Indus highway was selected which is one of the most prominent and strategic highway of the country. This highway serves as a hinterland connectivity, as it connects the ports of the country to its major cities

However, it is a very lengthy highway with a total length of 1264 km, connecting Karachi port city to Peshawar, therefore a section of it was considered for this study which is between Peshawar and Dera Ismail Khan. The length of this section is 302 km and passes through various rural and urban areas. The selected section was divided into five segments between the prominent cities on the route. The number of lanes in this segment is mostly two, however when it passes through urban areas, the number of lanes has been increased to four. The type of median varies from none to grassy, curbs and paved at places. It has a number of horizontal and vertical curves with various bridges and access points. Since the highway is passing through various rural and urban areas, the speed limits it has, are varying. Since it is a key source of goods transfer, the traffic through it has a huge portion of heavy vehicles.

The data about the crashes in a specific area are collected by the regional police as per their jurisdiction, which is in the form of a properly detailed lodged FIR (First Information Report). Therefore, this study is based on the data extracted from the individual FIRs of each crash accidents in a specific segment of road from the police station of concerned jurisdiction.

These FIRs provide details on the types of vehicles involved, number plates, details on the conditions of drivers in the accident having information about the age, addiction and route or duration of driving. It also provides information on the level of experience the involved drivers had, their license details and the information about the possible causes after investigation which could have initiated these accidents.

A total of 8 years' traffic data was taken from the year 2011 to 2018. Since crash number is a count variable, hence best analyzed through Poisson and Negative binomial regression models.

Since the length of road segment with two lanes is higher than road length with four lanes, therefore the number of crashes occurring at two lanes is also higher than that occurring at four lane section as represented in figure 1.

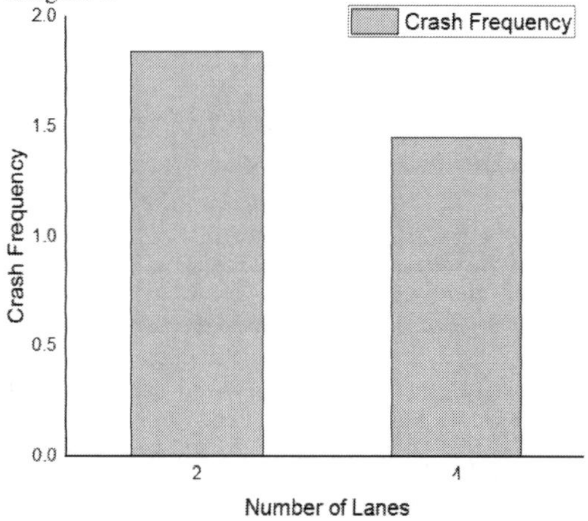

Figure 1. Graph between crash frequency and number of lanes.

The relationship between the crash frequency and speed limits at specific segments as per the statistical data under study shows that crashes at lower speed sections are more frequent than on high speed sections as depicted in figure 2.

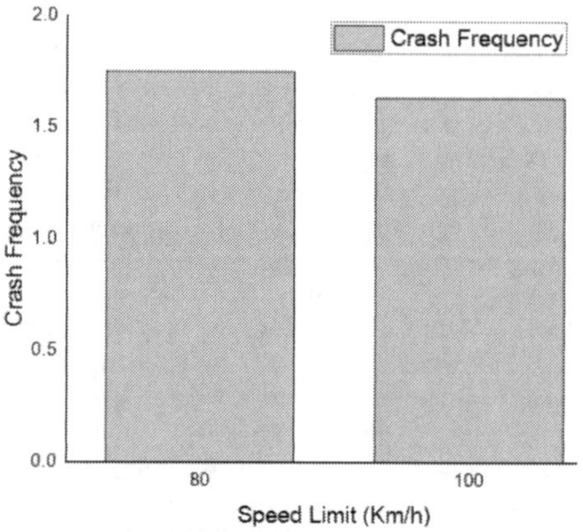

Figure 2. Graph between crash frequency and speed limits.

The pattern between the crash frequency and driver's behavior, climate effect, pavement condition and vehicle issues have been depicted in the figure 3.

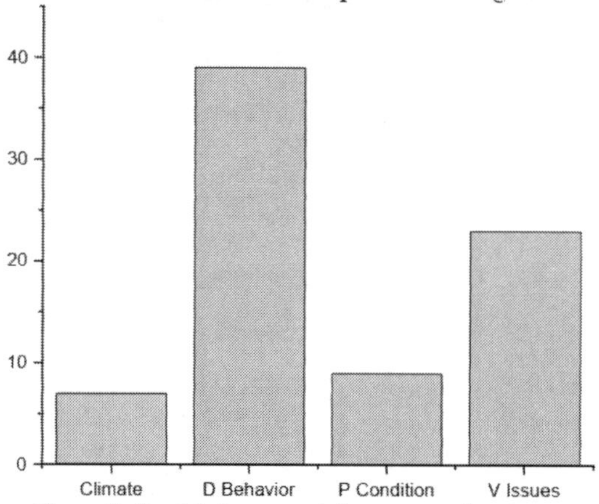

Figure 3. Graph between crash frequency and various factors.

However, the utility of Poisson regression model is limited by the ratio of mean to variance of the data which has to be equal to one. While this restriction is relaxed for the negative binomial regression. The data was found to be over-dispersed, the model suitable for this study is negative binomial regression (NBR). NBR actually is an addition of the Poisson regression, which have been provided with the measure for gamma distribution error [23]. Moreover, the selection of NBR was based on the performance of both dispersion and Vuong statistic parameters.

The mathematical expression to represent the negative binomial regression is given as;

$$Prob[Y = y_i \mid \varepsilon] = \frac{\exp[-\lambda_i \exp(\varepsilon)](\lambda_i)^{y_i}}{y_i!},\ y_i = 0,1,2,\ldots \quad \textbf{(1)}$$

$$\lambda_i = \exp(X_i\boldsymbol{\beta} + \boldsymbol{\varepsilon}_i) \text{ for the ith observation; } \textbf{(2)}$$

Where, λ_i is the projected mean, β is a vector of venerable strictures, x_i is the factors selected as independent variables, and $\exp(\varepsilon_i)$ is a gamma error term with mean 1.0 and variance α^2. The adding of this constraint permits the variance to vary from the mean:

$$\mathrm{Var}[y_i] = \mathrm{E}[y_i]\{1 + \boldsymbol{\alpha}\mathrm{E}[y_i]\} = \mathrm{E}[y_i] + \boldsymbol{\alpha}\mathrm{E}[y_i]^2 \quad \textbf{(3)}$$

The variable α, stands for the dispersion of the data, where if it gets equal to 0, the equation would become the Poisson regression model.

4 Results and discussions

In order to analyze the selected variables using NBR, first the mean, standard deviation, minimum and maximum values of the dependent and independent variables have been depicted in the table 1.

Table 1. Various statistical parameters of the selected variables

No	Variables explanation	Mean	Std dev	Min	Max
1	Dependent Variable: Crash frequency	1.49	2.03	0.0	12
2	Proportion of Unit trucks on a segment	10.19	2.79	5	14
3	Width of lanes in meters	3.59	0.0782	3.65	3.9
4	Segment wise lanes (Numbers)	2.30	0.391	2	4
5	Length of each segment	1.11	0.789	0.15	6
6	Access points in a segment	2.09	2.87	0.0	14
7	Urban and rural specification (1 and 0 respectively)	0.44	0.55	0.0	1.0
8	Allowed speed at each segment (km/hr.)	87.89	14.87	80	100
9	U-turns provided in a segment	0.93	1.05	0.0	5.0
10	Climate Impact	0.82	0.99	0.0	1
11	Priority Negligence	2.24	1.03	3	9
12	Indecorous reversing	1.34	0.82	5	12
13	Yielding negligence	4.31	1.37	7	14
14	Vehicle control failure while driving	10.81	2.43	0.2	5
15	Inappropriate turns	8.95	1.63	0.0	6
16	Wrong way or direction	1.23	0.73	0.3	1
17	Inappropriate Maneuvering	3.78	0.058	3.0	6.0
18	Light conditions	2.96	0.89	3	4.5
19	Pavement surface	4.51	0.95	0.5	3

The results obtained after running the NBR analysis have been shown in table 2. Reference to this table it could be seen that the number of single unit trucks are found to be significant with a t-value of 3.35 in causing highway crashes. This could be attributed to the reason that lighter trucks and vehicles have higher speeds as compared to the heavy trucks resulting in overtaking and increased crash risk. Similarly, the effect of a segment length on the crash frequency turns out to be significant with a t-stat value of 2.7. The results indicate that segments with longer lengths had contributed more to the occurrence of accidents as compared to shorter segment lengths. This could be attributed to the reason that a longer length of a road segment has potentially more exposure to the different traffic types and hindrances. These sections generally have higher number of U-turns, turns & horizontal curves and entrance roads. Due to the prevalence of similar road features, drivers have the inclination to do over speeding which also increase the crash risk.

Table 2. Results obtained from NBR analysis

Variables explanation	Estimated Coefficient	t-Stat	p Value
Constant	-16.03	-3.02	0.0004
Proportion of Unit trucks on a segment	0.19	3.35	0.0007
Climate Impact	0.54	1.27	0.0013
Segment lane numbers	0.89	2.90	0.0041
Indecorous reversing	0.62	1.59	0.5803
Access points in a segment	0.08	2.270	0.0244
Vehicle control loss while driving	0.28	1.55	0.0008
Inappropriate Maneuvering	0.60	2.44	0.0035
U-turns provided in a segment	0.08	1.04	0.3016
Width of lanes in meters	3.09	3.03	0.0021
Priority Negligence	0.78	2.54	0.0020
Length of each segment	0.29	2.70	0.0059
Yielding negligence	0.82	1.78	0.0033
Urban and rural specification (1 and 0 respectively)	0.69	1.58	0.2034
Inappropriate turns	0.72	1.63	0.0062
Wrong way or opposite direction	0.40	1.25	0.0006
Allowed speed at each segment (km/hr.)	0.02	1.21	0.2309
Light conditions	0.67	1.16	0.0051
Pavement surface	0.10	1.35	0.0064
Dispersion parameter (a)	0.51	4.709	0.0012

The climate effect on the occurrence of highway crashes is found significant with a t-value of 1.27, this is because in winters there is huge fog and smog observed on this section of road. It results in very poor visibility and consequently traffic jamming. Similarly, the loss of control on vehicle and inappropriate maneuvering with t-values of 1.55 and 2.44 respectively were also found significant. The loss of control on a vehicle yields in a very high crash risk. The careless and inappropriate maneuvering puzzles the other drivers creating unsafe crossing & braking distances and subsequently resulting in crashes.

The width of lanes also turns out to have significant effect on the occurrence of highway accidents with a t-value of 3.03. The results reveal that wider lanes have higher tendency of resulting in an accident as compared to the narrow or constricted lanes. Though wider lanes mean more space, decision making time and better visibility, but still the results indicate otherwise. The potential reason is that drivers on lanes wider than normal tends to increase their speed which enhances the crash risk. Moreover, in a wider lane the drivers would see space along vehicles ahead in the same lane and will tend to cross or accompany these vehicles resulting in increased endangerment. The drivers on narrow lanes remains more careful with their speed and overtaking behaviors resulting in reduced crash risk.

The recognition, acceptance and understanding of the priorities in traffic while driving at a highway plays a very critical role in crash causation. The negligence in lane changing, passing, turning and yielding priority can result in misunderstanding and subsequently result in higher probability of crash occurrence. The t-values for priority and yield negligence were found to be 2.54 and 1.78 consequently with both of these variables being significant. Furthermore, making inappropriate turns or turns with wrong indicator also significantly increased the crash frequency with a t-value of 1.63. Similarly, driving in the opposite direction is very dangerous and results in very high probability of head-on and sidewise collisions having a 1.25 t-value. This puts the other drivers in jeopardy of making wrong maneuvers and turns to avoid the crash and hence result in collision with other vehicles.

Moreover, the number of lanes are also found to be significant in their effect on the crash frequency. It is attributed to the reasons that increased number of lanes would consequently result in increased traffic volumes, lane changing and higher speeds. All these reasons ultimately create circumstances in which higher crashes takes place. Similarly, urban areas are found significant in increasing crash frequencies with a t-value of 1.58. This is due to the increased volume of traffic, higher number of pedestrians, cycles, motorcycles and various carts. Another reasons which causes accidents are the encroachments in the urban areas which results in traffic jams and consequently impatience in drivers, resulting in close overtakes and higher crash frequency.

The light conditions prevailing at the road and the surface conditions of the road are also found significant. The poor light conditions results in poor visibility and wrong judgment of the size, speed and position of the other vehicles. While the poor surface conditions can result in the loss of control over a vehicle and bursting, deterioration of tires and other parts consequently resulting in high vehicular crashes.

The speed limit, number of U-turns, number of access points and indecorous reversing though are very critical for traffic safety and crash frequency, but were found insignificant in the results of this study.

5 Conclusion

The number of traffic has increased manifold with the passage of time and so has the highway crashes. Highway crashes result in the loss of precious lives and property, hence enhancing the traffic safety is of remarkable prominence. The statistical data of the highway crashes and the factors due to which it occurs is best analyzed using Poisson or Negative Binomial Regression for reliable results.

Highway crashes have been a major concern in Pakistan for the authorities as it results in a significant annual life and monetary loss. Indus highway is one of the major trade routes of the country and its crash data have not been analyzed up till now as per the information of authors. The highway crash frequency data along its causation factors of this highway for a 302 km section from Peshawar to Dera Ismail Khan was taken and analyzed using NBR. The data was arranged as per the eighteen most important factors that affects the crash frequency and analyzed using SPSS software.

The results indicate that the crash frequency is highly subject to the length of specific road segments, number of lanes and the width of the lanes and an increase is observed in crash frequency with increase in these parameters. Similarly, the light, pavement and climate conditions also have a significant effect on highway crashes and the number of crashes increases as the condition of above parameters gets deteriorated. Similarly, the driver behavior, attitude and law abidance also plays a key role in the occurrence of highway crashes. Negligence, lower understanding & recognition, violation of rules and carelessness significantly increases the crash frequency.

References

1. M. R. R. Shaon, X. Qin, Z. Chen, and J. Zhang, "Exploration of contributing factors related to driver errors on highway segments," *Transportation Research Record*, vol. 2672, no. 38, pp. 22–34, 2018.
2. M. R. R. Shaon, X. Qin, A. P. Afghari, S. Washington, and M. M. Haque, "Incorporating behavioral variables into crash count prediction by severity: A multivariate multiple risk source approach," *Accident Analysis & Prevention*, vol. 129, pp. 277–288, 2019.
3. A. P. Afghari, M. M. Haque, and S. Washington, "Applying fractional split model to examine the effects of roadway geometric and traffic characteristics on speeding behavior," *Traffic injury prevention*, vol. 19, no. 8, pp. 860–866, 2018.
4. S. Washington and M. Haque, "On the commonly accepted assumptions regarding observed motor vehicle crash counts at transport system locations," 2013.

5. M. Shaon and R. Rahman, "Quantifying Rural Highway Safety Performance: Application of the Highway Safety Manual and Development of Mixed Distribution Statistical Models for Predicting Crash Frequency," 2015.
6. S. Box, "New Data from VTTI provides insight into cell phone use and driving distraction," *Virginia Tech Transportation Institute*, vol. 27, 2009.
7. M. R. Rahman Shaon and X. Qin, "Use of mixed distribution generalized linear models to quantify safety effects of rural roadway features," *Transportation Research Record*, vol. 2583, no. 1, pp. 134–141, 2016.
8. A. Montella and L. L. Imbriani, "Safety performance functions incorporating design consistency variables," *Accident Analysis & Prevention*, vol. 74, pp. 133–144, 2015.
9. I. Khan, A. Khan, F. Aziz, M. Islam, and S. Shafqat, "Factors associated with helmet use among motorcycle users in Karachi, Pakistan," *Academic emergency medicine*, vol. 15, no. 4, pp. 384–387, 2008.
10. I. Ahmed, T. Islam, G. Ali, and M. M. Nawaz, "Pillion riders' cloth related injuries and helmet wearing patterns: a study of Lahore, Pakistan," *International journal of injury control and safety promotion*, vol. 23, no. 4, pp. 388–394, 2016.
11. D. Shinar, *Traffic safety and human behavior*. Emerald Publishing Limited, 2017.
12. A. Montella, L. Colantuoni, and R. Lamberti, "Crash prediction models for rural motorways," *Transportation Research Record*, vol. 2083, no. 1, pp. 180–189, 2008.
13. P. C. Anastasopoulos, F. L. Mannering, V. N. Shankar, and J. E. Haddock, "A study of factors affecting highway accident rates using the random-parameters tobit model," *Accident Analysis & Prevention*, vol. 45, pp. 628–633, 2012.
14. L. K. Spainhour, D. Brill, J. O. Sobanjo, J. Wekezer, and P. V. Mtenga, "Evaluation of traffic crash fatality causes and effects: A study of fatal traffic crashes in Florida from 1998-2000 focusing on heavy truck crashes," 2005.
15. A. Ahmed, B. A. Khan, M. B. Khurshid, M. B. Khan, and A. Waheed, "Estimating national road crash fatalities using aggregate data," *International journal of injury control and safety promotion*, vol. 23, no. 3, pp. 249–254, 2016.
16. K. Jadaan, I. Al-Hyari, H. Naghawi, R. Ammourah, and Z. Al Nabulsi, "Traffic safety in Jordan: magnitude, cost and potential countermeasures," *J. of Traffic and Logistics Engineering*, vol. 1, no. 1, pp. 54–7, 2013.
17. P. M. Carter, C. R. Bingham, J. S. Zakrajsek, J. T. Shope, and T. B. Sayer, "Social norms and risk perception: Predictors of distracted driving behavior among novice adolescent drivers," *Journal of Adolescent Health*, vol. 54, no. 5, pp. S32–S41, 2014.
18. R. J. Schneider, A. Sanatizadeh, M. R. R. Shaon, Z. He, and X. Qin, "Exploratory analysis of driver yielding at low-speed, uncontrolled crosswalks in Milwaukee, Wisconsin," *Transportation research record*, vol. 2672, no. 35, pp. 21–32, 2018.
19. J. Moeckli and J. D. Lee, "The making of driving cultures," *Improving Traffic Safety Culture in the United States*, vol. 38, no. 2, pp. 185–192, 2007.
20. A. P. Afghari, S. Washington, M. M. Haque, and Z. Li, "A comprehensive joint econometric model of motor vehicle crashes arising from multiple sources of risk," *Analytic methods in accident research*, vol. 18, pp. 1–14, 2018.
21. M. Asbridge, J. A. Hayden, and J. L. Cartwright, "Acute cannabis consumption and motor vehicle collision risk: systematic review of observational studies and meta-analysis," *Bmj*, vol. 344, p. e536, 2012.
22. D. Lord and F. Mannering, "The statistical analysis of crash-frequency data: a review and assessment of methodological alternatives," *Transportation research part A: policy and practice*, vol. 44, no. 5, pp. 291–305, 2010.
23. W. H. Greene and C. Zhang, "Econometric analysis (Vol. 5)," *Upper Saddle River, NJ: PrenticeHall*, 2003.

https://doi.org/10.1051/matecconf/201929601006

The Development of Safety Performance Functions for Roundabouts in Amman, Jordan

Diana AL- Nabulsi[1, a] and KhairJadaan[2]

[1] *Research Assistant, Civil Engineering Department, University of Jordan*
[2]*Professor, Civil Engineering Department, University of Jordan*

Abstract. It is now well established that crash occurrences at roadway segments or intersections are associated with a large variety of factors. Safety Performance Functions (SPF) are statistical models developed to predict crash frequencies for various design variables.

In Jordan, almost half of all road crashes occur at intersections, mostly in urban areas. There is a significant number of roundabouts throughout Amman which is increasing without evaluating their safety performance. To assess safety benefits of this kind of intersections, transportation professionals need the powerful statistical tool; the SPF.

This study aims to develop SPFs for roundabouts in Amman. The models consider the crash frequency, traffic volume and geometric features of all the studied 20 roundabouts. The developed SPFs were statistically significant (R^2 = 0.91) .The findings of the study revealed that crash frequency has a strong relation with the AADT, roundabout entry angle-degrees, entry path radius, splitter radius, pedestrian crossing structure, inscribed diameter, central diameter, circulating width, entry width, number of circle legs. The developed SPFs are evaluated through a comparison with others from developed countries.

1. Introduction

Road traffic safety has become an intensively researched topic with the aim of enabling the road authorities to identify safety improvement projects more efficiently. According to statistics released by the Jordan Traffic Institute (JTI) Jordan had 685 fatal crashes in 2017 with 76% of these crashes occurring in Amman of which 35% occur at intersections (JTI 2017). These facts calls for the need to establish a goal of reducing traffic related fatalities and severe injuries along the Jordanian roadway network (segments or intersections). Achieving this goal requires developing statistical models, referred to as Safety Performance Functions (SPFs) which provide realistic predictions of crash frequencies as a function of traffic volume and roadway geometries over a roadway network.

The roundabout is a form of intersection control which, if properly designed, can provide safety benefits when compared to stop and signal controlled intersections. 32 % of Amman's intersection crashes occur at roundabouts. Therefore, roundabouts have become increasingly popular in recent years leading to an increase in research into their effectiveness and safety performance.

This study provides an objective and unbiased method of measuring future safety for Amman roundabouts by developing SPFs to predict their crash frequencies. Comparing the predicted and observed crash frequency at a particular roundabout determines its relative safety.

2. Literature review

The first version of Highway Safety Manual (HSM) appeared in 2010 [1] , and it was a rescue to solve traffic safety issues on roads. The American Association of State Highway and Transportation Officials published it to give decision makers an idea on how to develop safety by anticipating the number of crashes depending on the road or intersection features (Peden, et al., 2004), then the SPF appeared. Road traffic crashes are discrete and nonnegative. They are usually over-dispersed with their variance higher than mean. Therefore, in order to account for over-dispersion in the fitted models, negative binomial (NB) distribution is typically adopted (Srinivasan, et al. 2011). The general functional form of an SPF: $N = \exp(\alpha+\beta\times \ln AADT + \gamma 1X1+ \gamma 2X2+\ldots+\gamma nXn)$ Where: N: predicted average crash frequency, AADT: annual average daily traffic, X1, X2, …, Xn = n roadway

[a] Corresponding author: diana_alnabulsi90@hotmail.com

geometric characteristics, and α, β, γ1, γ2, …, γn = regression coefficients.

After that, the idyllic methods to develop the SPF models were Poisson regression and Binominal regression [2]. The negative binominal regression which is in the HSM adaptive method, was superior, since it doesn't have any limitations and its results are over dispersion, as Srinivasan, et al. (2011) described it and showed that standard errors of the parameter estimates to be tend understated [3] . A roundabout is a form of intersections and it doesn't have to be circular. It aims to control traffic without the need for stop signs [4]. The modern roundabout is a type of intersection that indirectly provides traffic control without the use of stop signs or traffic signals. These roundabouts, if properly designed, can provide safety and traffic flow benefits when compared to stop controlled and signal controlled intersections [5]. Montella (2011) conducted a study on 15 roundabouts in Italy and identified the crash contributory factors at roundabouts by using association rule analysis. Crashes that occurred could be attributed to either a single or multiple factors. Among 62 different contributory factors, geometric design was found to be the most frequent factor, occurring in almost 60 percent of total crashes. Markings and pavement were also major contributory factors that led to crashes. Several studies in the United States have indicated that roundabouts are a helpful means to reduce frequency and severity of crashes [6] . Isebrands, et al., (2012) carried out a study related to higher-speed rural roundabouts with stop control before the conversion and observed a reduction of 67 percent of total crashes and a reduction of 87 percent of injury crashes [7] . Persaud, et al., (2001) also reported that converting traditional intersections from stop sign or signal control decreased all crashes by 35 to 72 percent and decreased injury crashes by 74 to 88 percent [8]. Daniels, et al., (2010) explained that the safety performance of an intersection or a segment of roadway is identified through developing a relationship that includes all potential variables. They proved that crash frequency is not only related to the annual average daily volume (AADT) but is also affected, at least partly, by the geometric features of roundabouts [9].

3. Methodology

The study is carried out according to the following steps as shown below:

A. Select 20 roundabouts in Amman characterized with a high volume since previous research identified the AADT as the variable with the greatest impact on crash prediction using SPF.
B. Collection of comprehensive information for the selected roundabouts. This step involves the acquisition of the crash and geometric data. The crash data for each roundabout were provided by the Central Traffic Department and the geometric data are extracted from Google Earth.
C. Conducting a statistical analysis of the collected data to identify the relationship between the predicted numbers of crashes and a set of geometric and traffic characteristics. Data are compiled and analyzed using both Excel spreadsheets and SPSS IBM V19.
D. Developing SPF models that presents the crash frequency as a function of AADT and other related probable geometric variables

4. Traffic and Geometric Characteristics of the Studied Sites

The safety situation at the studied sites may be judged by the actual number of car crashes that have occurred over time. Fig.1 shows the distribution of the observed crashes over four-year period (2015 – 2018). It can be seen that most of the roundabouts (about 70%) experienced an average of about 40 crashes per year. The descriptive statistics of the count of the studied roundabouts are shown in Table 1:

Table 1: Descriptive statistics for the analyzed (N=20) roundabouts

	Frequency	Percentage
Presence roundabout structure		
No	6	30
Yes	14	70
Presence of random police patrol		
No	4	20
Yes	16	80
Presence of pedestrian crossing structure		
No	12	60
Yes	8	40
Entry path angle , mean (SD)		28.90 (5.5)
Entry path radius, mean (SD)		69.5 (26.4)
Splitter island radius, mean (SD)		77.9 (29.2)
Roundabout central diameter (meters), mean (SD)		58.3 (29.6)

Roundabout inscribed diameter (meters), mean (SD)		74.05 (28.7)
Roundabout circular width diameter (meters), mean (SD)		8.9 (2.47)
Number of roundabout legs, mean (SD)		4.4 (0.6), median=4
Number of circle legs categorized		
<=4 legs	14	70
>4 legs	6	30
Roundabout leg Entry width (meters)		7.15 (3.6)
Annual average daily traffic Rate, mean (SD)		
AADT 2015		96931 (42230)
AADT 2016		100687 (44002)
AADT 2017		104730 (45824)
AADT 2018		108924 (47614)
Roundabout annual car crashes		
Total number of car crashes-2015, mean (SD)		129.55 (81.8)
Total number of car crashes-2016 , mean (SD)		165.05 (109.7)
Total number of car crashes-2017 , mean (SD)		179.30 (131)
Total number of car crashes-2018, mean (SD)	Frequency	188.05 (142.5)

The above table shows that 70% of the studied roundabouts had structures like bridges and pedestrian crossing features whereas 30% had no such structures. 80% of the roundabouts, had a regular police patrols stationed on the roundabouts that control traffic on daily basis, whereas 20 % had no police patrols. Assessing the analyzed bridges for the presence or absence of pedestrian crossing structures shows that:

60% had no crossing structures, whereas 40% had a built crossing structure. The mean entry path radius was equal to 69.5 + 26.4 meters. The mean distance for the bridges splitter island is calipered of 77.9 + 29.2 meters. The mean roundabout central diameter for the analyzed circles was 58.3 + 29.6 meters, whereas the mean inscribed diameter for the circles (meters) was 74.1 + 28.7, with a circular roundabouts diameter of 8.9 meters, SD = 2.5 meters.

70% of the studied roundabouts had four legs (exists), or less, and 30% have more than four legs, however, the mean legs number is centered at 4.4 + 0.6 legs. The mean roundabouts entry width was equal 7.15 + 3.6. With regards to the measured annual car crashes, they suggest increasing trend:

The mean number of car crashes rises from 129.6 in 2015 to 165.1 in 2016, then it continued to rise between the years 2017 and 2018.

5. Model Development

The Generalized Mixed Methods R-M Analysis assumed the data of roundabouts across years are correlated, likewise for certain characteristics like the AADT, due to the presence of strong correlations between roundabout's repeated car crash measures. On this basis, a fixed effects model was requested, and the program was allowed to model the variance components between the twenty roundabouts, assuming that each roundabout may have different variances and co-variances (rise and fall) of their annual car crash crash's counts across the vector of time (years).

The variance component was also added in order to enhance the fit of the model via allowing each of the roundabouts to differ in their trajectories due to other factors inherent to their nature since they're geometrically different roundabouts.

The Generalized Mixed Methods requested modeling was a Negative binomial with logarithmic link due to the inherent skewness of car crash count variable, which follows the Poisson distribution family.

The model fit was assessed using the Baysian (AIC and BIC criteria).The over dispersion was assessed using methods like Scatterplots of observes versus predicted plots and predicted versus residual plots. Table 2, on the roundabouts measured annual crash rates across the four years between (2015-2018) for each roundabout.

 https://doi.org/10.1051/matecconf/201929601006

Table-2: Multivariate Mixed effects Generalized Negative Binomial Regression Analysis (N=20 roundabouts).

					95% C.I Exp (Beta)		
	Beta coefficient	Std. Err.	t-value	Exponentiated (Beta)	Lower	Upper	p-value
(Intercept)	-3.150	2.200	-1.430	0.043	0.001	3.480	0.157
(AADT/1000) Score	0.011	0.002	6.290	1.011	1.007	1.014	<0.001
Roundabout entry angle-degrees	0.114	0.043	2.630	1.210	1.028	1.220	0.011
Entry path radius meters	-0.022	0.035	-0.642	0.978	0.912	1.050	0.523
Splitter radius – meters	0.047	0.060	0.785	1.048	0.930	1.182	0.435
Pedestrian crossing structure = yes present	0.124	0.137	0.908	1.131	0.862	1.490	0.367
Inscribed diameter	0.051	0.060	0.866	1.052	0.936	1.184	0.390
Central diameter=meters	-0.033	0.013	-2.540	0.097	0.943	0.993	0.013
Circulating width	-0.023	0.080	-0.290	0.977	0.832	1.147	0.773
Entry width-meters	-0.011	0.070	-1.560	0.897	0.781	1.031	0.124
Number of roundabout Legs<=4	0.577	0.198	2.914	1.781	1.781	1.199	0.005

The independent variables affecting the number of predicted crashes were found to include the Annual Average Daily Traffic (AADT) in addition to some geometrical features of the roundabout. The latter include roundabout entry angle, entry path radius splitter radius, presence or absence of pedestrians crossing structure, inscribed, central and circulating as well as entry width measures).

The model was specified via setting the time (years) as repeated measure, the subjects were the eighty records of individual yearly roundabouts measures. The roundabouts themselves were set as random factors and we requested the program to estimate a variance components model. The model was found to have the following form:

N = eβ0 * AADTβ1 * **e (β2X2 + β3X3 + …….)**
Where:

X1: Roundabout entry Angle-degrees, X2: Entry path radius-meters, X3: Splitter radius-meters. X4: Pedestrian crossing structure = yes present, X5: Inscribed diameter, X6: Central diameter (meters), X7: Circulating width, X8: Entry width-meters, X9: Number of circle legs (4 legs are used), (β_0, β_a): Beta coefficient from table 2.

The results showed the analysis model with the Generalized Mixed methods indicated that the for 4-legs roundabout or less annual AADT, entry angle, correlated significantly but positively with higher crash rate growth. Whereas the RA central diameter correlated with lower crash rates, i.e., the correlation was significant and negative when accounting for the other predictors in the analysis ($p = 0.013$) a series parameters was used to evaluate and to determine how strong the statistical analysis was, such as AIC, BIC, AICC and BICC.

The model was validated by comparing the actual (Observed) and the predicted annual crashes producing a statistically significant relationship ($R^2 = 0.91$) as shown in fig.2

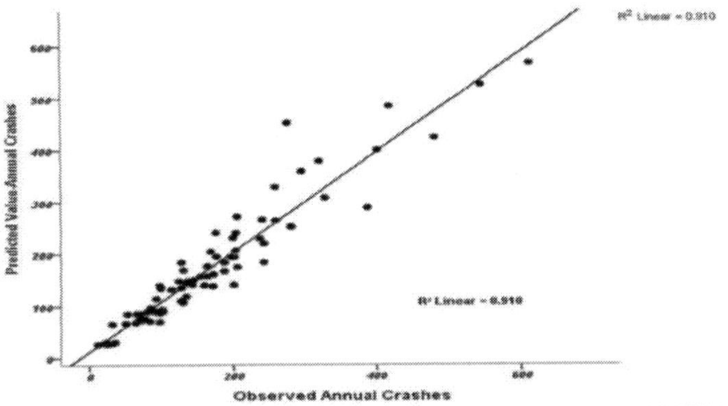

Figure 2: scatterplot of observed versus model predicted annual crashes

6. Conclusions

SPFs have been used and proved to be effective safety measures in many countries but not yet implemented in Jordan. Developing SPFs using local data will provide more reliable crash frequency estimates than applying any calibration procedure.

This study involved the estimation of safety performance for roundabouts in Amman, through the development of SPFs to estimate the number of annual crashes at any given roundabout and to determine the impacts associated with various traffic and geometric characteristics,

However, the developed model tend to be reactionary and the approach of using only AADT has the limitation of being unable to identify the reason for a crash which requires a more proactive model. Therefore, a multivariate analysis was performed to determine how a variety of variables can contribute to crashes.

The results showed that the for 4-legs or less roundabouts, the factors of AADT and entry angle, correlate significantly and positively with higher crash rate, whereas the roundabout central diameter correlated with lower crash rates, i.e., the correlation was significant and negative when accounting for the other predictors in the analysis ($p = 0.013$)

References

1. American Association Of State Highway and Transportation Officials (AASHTO), (2010). **Highway Safety Manual**, 1st Edition, Washington, DC.
2. Peden M, Scurfield R, Sleet D, Mohan D, Hyder A, Jarawan E, Mathers C, (2004). **The world report on road traffic injury prevention. WHO, Geneva.**
3. Srinivasan, R., Haas, P., Dhakar, N., Hormel, R., Torbic, D. and Harwood, D., (2011), **Development and Calibration of Highway Safety Manual Equations for Florida Conditions.** Florida Department of Transportation.
4. Quezon T.E., Kumala T., and Aynetaw B., (2017), **Evaluation on the Operational Characteristics of Gerji-Imperial Roundabout: A Case Study in Addis Ababa.**
5. McIntosh k. (2011).**Evaluating the Performance and Safety Effectiveness of Roundabouts**, Opus International Consultants Inc. The Michigan Department of Transportation, USA.
6. Montella, A. (2011). **Identifying crash contributory factors at urban roundabouts and using association rules to explore their relationships to different crash types.** Crash Analysis & Prevention, 43(4), 1451-1463
7. Isebrands, H., & Hallmark, S. (2012). **Statistical analysis and development of crash prediction model for roundabouts on high-speed rural roadways**. Transportation Research Record, 2312(1), 3-13.
8. Lyon C., Persaud B., Hahn J. (2011). **Safety Performance Functions for Ramp Terminals at Diamond Interchanges** (Report No. CDOT 2011-9). Colorado Department of Transportation, Denver, Colorado.
9. Daniels, S., Brijs, T., Nuyts, E., & Wets, G.(2010). "**Explaining variation in safety Performance of roundabouts**" Crash Analysis & Prevention, Vol.42, No.2, Elsevier, pp. 393-402.

How Connected Autonomous Vehicles Would Affect Our World? —— A Literature Review on the Impacts of CAV on Road Capacity, Environment and Public Attitude

Shuya Zong[1, a]

[1]*Department of Civil and Environment Engineering, Hong Kong Polytechnic University*

Abstract. With the rapid development of technology, connected autonomous vehicle (CAV) is getting close to the reality. The application of CAV causes changes to road capacity, gas emission, public attitude and other realms. Lots of efforts have been spent in quantifying the potential changes and this paper is an attempt to review the relevant researches. There will be three sections, presenting review of the impacts on road capacity, environment and public attitude respectively. There is a large amount of papers making models to predict future road capacity with various penetration rate of CAV and they obtain quite different interesting results. To predict the future condition more properly, more stochastic models should be proposed. In terms of influence on environment, it may be hard to conclude whether CAV will exacerbate or relieve global warming by looking at current researches. It would be valuable to conduct a quantitative analysis on this issue. For the public attitude, this paper mainly focus on whether people are willing to use CAV and future efforts that may help with the promotion of CAV.

1 Introduction

Connected and automated vehicle (CAV) has aroused interest in the academic field as well as in automobile companies. It is predicted that the amount of connected cars sold will become more than 152 million in the world by 2020 [1]. 20-40% of vehicles are expected to be automated by 2030 and the full penetration could be realistic in few decades [2]. CAV is featured by its easier access to timely information, strong data processing ability and precision control. Due to these features of CAV, the reaction time and the distance between vehicles can be reduced significantly. As a result, the application of CAV increases driving safety, comfort and potential revolution in transportation framework. Understanding the impacts of CAV will be beneficial to future transportation control, policy making and urban design. Considerable research efforts have been devoted to the impact of CAV on road capacity, traffic jam, environment and public attitude. By using different analytical tools and data collection approaches, they provide interesting anticipation of the future of CAV application. This paper will attempt to review the relevant research to capture the potential influence of CAV on the road capacity, environment and public attitude. This paper will also propose the future research potential by looking at the limitations of current researches.

There will be three sections in this paper. The first chapter discusses the influence of road throughput with different penetration rate of CAV. The second chapter then explore how environment will be affected by CAV and the third chapter asks if the public will accept CAV with positive responses. Gaps and limitations of current work will also be discussed in each chapter, providing ideas for future researches.

2 Impacts on road capacity

Emerging CAV technologies have tremendous potential to improve traffic efficiency, safe and stability. To further apply the technologies in traffic control and planning, it is critical to evaluate how the road capacity would be improved with increasing penetration rate of CAV [3]. The research methods are either setting theoretical framework of road capacity (e.g. [4], [5]) or making simulation of CAV penetrated flow (e.g. [6], [7]).

A theoretical formulation for physical lane capacity of a single-lane road considering AV (automated vehicle) penetration rate, vehicle spacing and platoon size is derived by Chen et al in [4]. It was then found that a strict segregation of AV and RV (RV: regular vehicle) would result in lower capacity while mixed-use would realize higher capacities. The results can be illustrated in the figure 1 below. The formulation is extended further to multi-lane highway and it was found that AV should be used in most efficient lanes to maximum the road capacity. Chen et al developed a mixed routing equilibrium model to describe the flow of AV and CV with the presence of AV zones in [4]. They showed that an optimal deployment of AV zone should be applied to maximize the positive effect of AV and CV on road

[a] Corresponding author: sue.zong@connect.polyu.hk

capacity. Based on [4], Ghiasi in [8] proposed an analytical framework with a Markov chain representation of spatial distribution of heterogeneous headway on the highway. This framework provides a useful tool to examine the capacity change under different autonomous vehicle penetration.

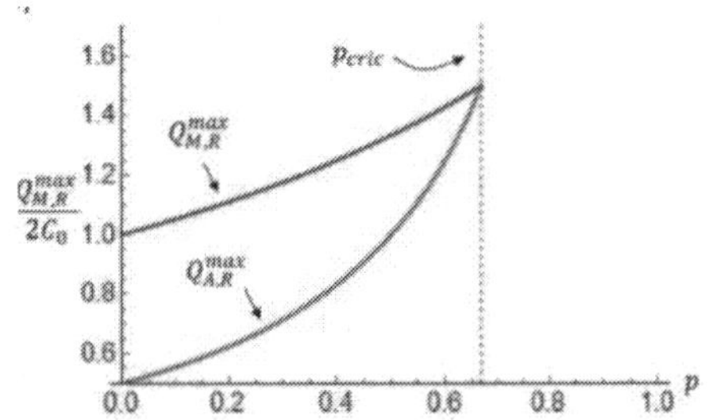

Figure 1. Mixed-use of AV and RV realizes higher capacity. $Q^{max}_{A,R}$:maximum capacity under the policy where AV and RV are separated. $Q^{max}_{M,R}$:maximum capacity under mixed use. P: AV penetration rate[4]

Apart from theoretical research, there are papers modelling the traffic flow after penetration of CAV(e.g. [9], [10], [11]). The change in road capacity depends highly on the proportion of autonomous cars: Levin and Boyles in [9] found that the proportion of AV should not exceed 75% to avoid the negative effects of inefficiency caused by mixed traffic based on a shared intersection model. Talebpour and Mahmassani in [10] analysed the effect of autonomous vehicles and connected vehicles individually and suggested that the throughput will increase if the penetration rate is medium. Moreover, they come up with a model describing the combined effects of those two types of cars. The results is explicitly illustrated in the figure 2 below.

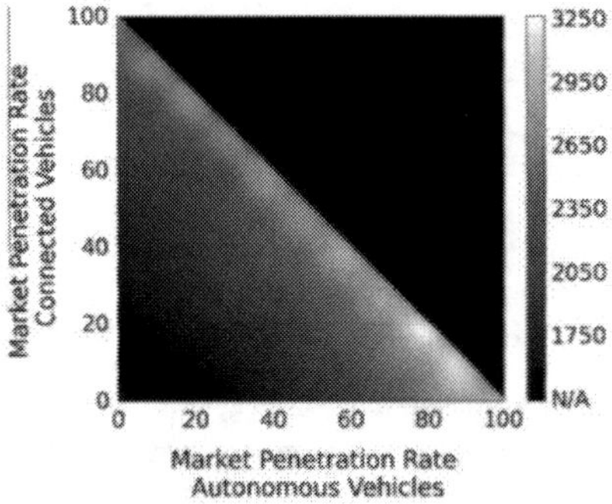

Figure 2 . Maximum throughput (veh/h/lane) at different market penetration rates of connected and autonomous vehicles for a platoon of regular, connected, and autonomous vehicles with infinite length [10].

The increase in road capacity needs to be quantitatively investigated. The increase in capacity is expected to be negligible (1%-4%) if the CAV do not cooperate [12] and to be higher that 400% if efficient cooperation is achieved [13]. Lioris et al in [14] proved that connected vehicles can double or even triple the intersection capacity by simulation by using a memory-less queuing model with an on/off service process to simulate the behavior of CAV at an urban intersection and a case study in Los Angeles. However, it should be noticed that they choose the increase rate of demand and service (T) based on the empirical experience in a scenario where all vehicles are connected. As it is unrealistic to assume a fully connected condition at present, further research should be done to evaluate the value of T under different penetration rate of CAV. It is noticeable that most of the models proposed at present are not applicable for network of intersections. However, the road system in an urban area is often complex and contains various of intersections connected together. Therefore, it would be valuable to put effort in the model simulating more stochastic scenarios.

3 Impacts on Environment

The impact on environment is not the core consideration in the CAV development, but the influence could be significantly[15]. By introducing changes in travel demand, vehicle design and car following profiles, CAV will affect road energy consumption and greenhouse gas(GHG) emissions positively or negatively [16]. It is worth to notice that vehicles which are only automated but not connected may not be able to reduce energy consumption. A simulation found that an increase of energy consumption of an autonomous vehicle is 3% more than the traditional vehicle due to its following algorithms[17].In comparison, connected vehicles are better at predicting the behavior of preceding car by precise sensing ability and they can adjust their headway at a smooth motion. Wadud et al in [16] apply the ASIF framework proposed by Schipper in 2002 [18] to organize the potential mechanisms through which CAV affect the environment. They analysed the potential impacts in various illustrative scenarios and realized that the automation of vehicles might reduce the GHG emissions by nearly 50% or double it depending on which condition dominates.

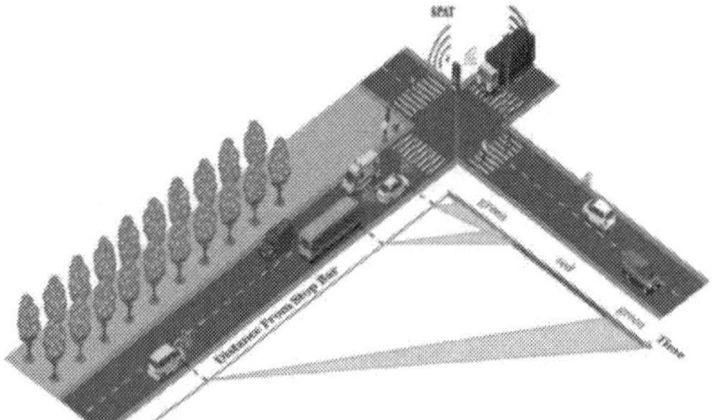

Figure 3. Schematic of eco-driving with SPAT preview [19]

As shown in the figure 3 above, The technology of Vehicle-to-Infrastructure(V2I) connectivity enables the Signal Phase and Timing (SPaT) to be broadcast to approaching vehicles, thus the vehicles are able to adjust their speed to make themselves arrive at the intersection at green light [19]. The models generated several researchers have supported the positive results (e.g. [20], [21]). However, it is noticeable that in the mixed traffic flow, the timely brake of CAV may be disruptive for the traffic environment. The micro models proposed is not sufficient to analyse this scenario and a smoother velocity profile should be developed considering the comfort of passengers and human drivers. Rakha and Kamalanathsarma in [22] investigated into this problem and by using a multi-stage dynamic model, they provided the speed trajectory which can save 19% energy.

How cars follow each other affects the fuel consumption significantly. Few branch of literature focus on the possible reduction in fuel use due to change in car following profiles. Li and Peng in [23] showed that around 30% energy can be saved by adopting a fuel-optimal following strategy. A driver-vehicle-infrastructure cooperative framework which can save more than 20% energy is proposed by Qi in 2016 for electric vehicles [24]. Though framework is not designed for CAV, the principle is valuable. The complex influence of communication delay is discussed by Ge and Orosz[25].

Though there have been papers analysing effect of CAV quantitatively, few of them focus on the quantitative impact on environment. It would be valuable to conduct more simulations to quantify the effect on environment based on new car-following profile with different penetration rate of CAV in the future. Moreover, even though vehicles are connected, how an automated vehicle and a human driver process the data of traffic environment can be significantly different. More research should be done to investigate how this difference would affect the car following and therefore the GHG emission.

4 Impacts on Public Attitude

Driving is a social activity and with the development of CAV technology, some social problems raised as well. Though it has been predicted that CAV will enable new user groups including children, disabled, elderly to enjoy the mobility, it does not mean that the society would be enthusiastic to CAV. Therefore, it is essential to monitor the public concerns on this technological development [26]. Several researches (e.g. [27], [28], [29]) have been conducted, trying to answer related questions: What will be the attitude of public towards CAV? Will people be willing to buy CAV? Will drivers be comfortable to share the road with CAV? The research in this area is multidisciplinary and usually involve psychological or social principles. These researches in a way suggest to which degree can public accept new advanced technology and the results therefore can be used in sociological fields. Tennant et al. conduct a cross national survey of drivers in 11 European countries with sample size of 1000 in each country in [30]. In this survey, more respondents give less enthusiastic attitude towards using CAV or driving alongside CAV and drivers who describe driving as an enjoyable activity tended to be more negative about AV. These results are consistent with the research conducted by European Commission in 2017 [31]. In terms of the willingness to use the CAV, the pricing of CAV plays an important role. 37% of the respondents in US indicated that they ‘definitely would’ or ‘probably would’ purchase a CAV as their next vehicle before learning the estimated price of $3000 while this percentage dropped to 20% after they are told the prices [32]. The adoption of shared autonomous vehicles changed obviously under different pricing scenarios according to a survey conducted in US [33].

The problem of willingness of sharing road with CAV has not been given enough attention. It seems that there is lots of consideration on how autonomous vehicles would interact with human driven cars (as shown in the first section in this paper), while how human will response to CAV has not been studied thoroughly. It is always assumed as default that the interaction of human with autonomous vehicle is similar to interaction with traditional cars [30].

No matter how advanced the CAV technology has been, the benefits associated with CAV can not be achieved unless the public use, pay and buy CAV. One of the major concerns related to CAV is how to promote them and how well do people understand the autonomous vehicles. Moreover, CAV are not all fully automated and there are 6 levels of autonomous vehicles (Level 0 to Level 5) according to SAE [34]. It is predictable that there will be a long transition period when autonomous vehicles of heterogeneous levels share the road [35]. Therefore, several concerns arise: Do drivers understand their driving tasks with different levels of CAV? Do consumers hold a clear understanding of current taxonomy of CAV? These concerns are quite significant to the promotion and safety of CAV. The confusion among consumers might lead to safety issues like car accidents. A branch of papers have indicated that some of the names of level 2 autonomous vehicles which are available now make drivers over confident in cars [36,37]. For instance, the name ‘Autopilot’ used by Tesla makes considerable number of consumers to believe that they can take a nap when driving, which is very dangerous [36]. Speelt et. al. suggested that prevailing classification system of CAV is misleading to some degree and they thus created a new classification method which contained only two levels [38]. However, a two-level classification seems unrealistic as the autonomous driving system developed by car companies contain various levels of automation. Therefore, how to break consumers' confusion is an interesting topic in the future CAV promotion.

5 Conclusion

It is predictable that autonomous vehicles would be on our highways and streets in few years, making difference to the performance of our transportation network. The issues associated with the operation of the new transportation system will be challenging and comprehensive. From pure mathematical framework, to models incorporating real world conditions, to social surveys, multiple methods of research have been used to anticipate the impacts of CAV. By reviewing literature related to the influence on road capacity, environment and public attitude, it is found that the complexity of these issues is huge. Though a general positive future is anticipated by most academics, it is hard to give a conclusion on either how much the road capacity would change or how the GHG emission would change with penetration of CAV. More efforts should be put in modelling network of intersections and changes of human drivers after introducing CAV to analyse the potential impacts quantitatively. It is hard to say that the public hold a very positive attitude towards CAV based on current surveys. To promote the CAV among the public, a simpler classification system of AV might be helpful. Policy makers should also pay attention to the public attitude and role of responsibility for human drivers with various levels of CAV to adjust the business or road control policies.

References

1. McCarthy, N.(2015). Connected cars by the numbers. Forbes.
2. Litman, T. (2017). Autonomous vehicle implementation predictions. Victoria Transport Policy Institute.
3. Litman, T., 2015. Autonomous vehicle implementation predictions implications for transport planning. Transportation Research Board 94th Annual Meeting.
4. Chen, D., Ahn, S., Chitturi, M., & Noyce, D. (2017). Towards vehicle automation: Roadway capacity formulation for traffic mixed with regular and automated vehicles. Transportation Research. Part B, Methodological, 100, 196.
5. Gong, Shen & Du, 2016. Constrained optimization and distributed computation based car following control of a connected and autonomous vehicle platoon. Transportation Research Part B, 94, pp.314–334.
6. Talebpour, & Mahmassani. (2016). Influence of connected and autonomous vehicles on traffic flow stability and throughput. Transportation Research Part C,71, 143-163.
7. Young, S.H. & The George Washington University. Engineering Mgt Systems Engineering., degree granting institution, 2016. A Model-Based Framework for Predicting Autonomous Unmanned Ground Vehicle System Performance.
8. Ghiasi et al., 2017. A mixed traffic capacity analysis and lane management model for connected automated vehicles: A Markov chain method. Transportation Research Part B, 106, pp.266 - 292.
9. Levin, & Boyles. (2016). A multiclass cell transmission model for shared human and autonomous vehicle roads. Transportation Research Part C, 62, 103-116.
10. Yoon, J.-woo & Kim, B.-woo, 2016. Vehicle position estimation using nonlinear tire model for autonomous vehicle. Journal of Mechanical Science and Technology, 30(8), pp.3461–3468.
11. Shladover, S., Su, D., & Lu, X. (2016). Impacts of Cooperative Adaptive Cruise Control on Freeway Traffic Flow. Transportation Research Record, 2324(2324), 63-70.
12. Fernandes, P., & Nunes, U. (2012). Platooning With IVC-Enabled Autonomous Vehicles: Strategies to Mitigate Communication Delays, Improve Safety and Traffic Flow. IEEE Transactions on Intelligent Transportation Systems, 13(1), 91-106.
13. Lioris, Pedarsani, Tascikaraoglu, & Varaiya. (2017). Platoons of connected vehicles can double throughput in urban roads. Transportation Research Part C, 77(C), 292-305.
14. Vahidi, & Sciarretta. (2018). Energy saving potentials of connected and automated vehicles. Transportation Research Part C, 95, 822-843.
15. Wadud, Z., MacKenzie, D., & Leiby, P. (2016). Help or hindrance? The travel, energy and carbon impacts of highly automated vehicles. Transportation Research, Part A, 86, 1-18.
16. Mersky, & Samaras. (2018). Corrigendum to "Fuel economy testing of autonomous vehicles" [Transp. Res. Part C: Emerging Technol. 65 (2016) 31–48]. Transportation Research Part C, 87, 212-217.
17. Schipper, L. (2002). Sustainable urban transport in the 21st century. A new agenda. Transportation Research Record, (1792), 12-19.
18. Miller, S.A. & Heard, B.R., 2016. The Environmental Impact of Autonomous Vehicles Depends on Adoption Patterns. Environmental science & technology, 50(12), pp.6119–6121.
19. Mahler, G., & Vahidi, A. (2014). An Optimal Velocity-Planning Scheme for Vehicle Energy Efficiency Through Probabilistic Prediction of Traffic-Signal Timing. IEEE Transactions on Intelligent Transportation Systems, 15(6), 2516-2523.
20. Rakha, Ahn, Moran, Saerens, & Bulck. (2011). Virginia Tech Comprehensive Power-Based Fuel Consumption Model: Model development and testing. Transportation Research Part D, 16(7), 492-503.
21. Kamalanathsharma, R., & Rakha, H. (2013). Multi-stage dynamic programming algorithm for eco-speed control at traffic signalized intersections. 16th International IEEE Conference on Intelligent Transportation Systems (ITSC 2013), 2094-2099.
22. Li, S., & Huei Peng. (2011). Strategies to minimize fuel consumption of passenger cars during car-following scenarios. Proceedings of the 2011 American Control Conference, 2107-2112.
23. Qi, X. (2016). Next Generation Intelligent Driver-Vehicle-Infrastructure Cooperative System for Energy Efficient Driving in Connected Vehicle Environment. 214.
24. Ge, & Orosz. (2014). Dynamics of connected vehicle systems with delayed acceleration feedback. Transportation Research Part C, 46(C), 46-64.
25. Becker, F., & Axhausen, K. (2017). Literature review on surveys investigating the acceptance of automated vehicles. Transportation, 44(6), 1293-1306.
26. Lamotte, R., de Palma, A. & Geroliminis, N., 2017. On the use of reservation-based autonomous vehicles for demand management. Transportation Research. Part B, Methodological, 99, p.205.
27. Grush, B. & Niles, J., 2018. The end of driving : transportation systems and public policy planning

for autonomous vehicles, Amsterdam, Netherlands: Elsevier.

28. Hudson, Orviska & Hunady, 2019. People ' s attitudes to autonomous vehicles. Transportation Research Part A, 121, pp.164 – 176.
29. Schipper, L. (2002). Sustainable urban transport in the 21st century. A new agenda. Transportation Research Record, (1792), 12-19.
30. European Commission. (2017). Special Eurobarometer 460: Attitudes towards the impact of digitisation and automation on daily life.
31. J.D.Power.(2012). Vehicle Owners Show Willingness to Spend on Automotive Infotainment Features, Westlake Village.
32. Banal, Kockelman, & Singh. (2016). Assessing public opinions of and interest in new vehicle technologies: An Austin perspective. Transportation Research Part C, 67(C), 1-14.
33. SAE International. 2019. Taxonomy and Definitions for Terms Related to Driving Automation Systems for On-Road Motor Vehicles. [ONLINE] Available at: https://www.sae.org/standards/content/j3016_201806/. [Accessed 20 July 2019].
34. Prakah-Asante, K.O., Strumolo, G.S. & Curry, R., 2018.
35. IIHS. 2019. New studies highlight driver confusion about automated systems. [ONLINE] Available at: https://www.iihs.org/news/detail/new-studies-highlight-driver-confusion-about-automated-systems. [Accessed 20 July 2019].
36. IIHS. 2019. What ' s in a name? Drivers ' perceptions of the use of five SAE Level 2 driving automation systems. [ONLINE] Available at: https://www.iihs.org/topics/bibliography/ref/2190. [Accessed 20 July 2019].
37. 10th International Driving Symposium on Human Factors in Driver Assessment, Training, and Vehicle Design. 2019. Consumer confusion with levels of vehicle automation.[ONLINE]Available at: https://drivingassessment.uiowa.edu/sites/drivingassessment.uiowa.edu/files/wysiwyg_uploads/combinedprogramschedulefiles_6_10_19_0.pdf. [Accessed 20 July 2019].

Study on the Urban Rail Transit Sleeper Spacing Considering Vehicle System

Yu Zou[1], Yongpeng Wen[1, a] and Qian Sun[1]

1 School of Urban Railway Transportation, Shanghai University of Engineering Science, 333 Long Teng Road, Shanghai 201620, China

Abstract. To design the optimal sleeper spacing of the track and reduce the vertical vibration of the rail, the influence of the sleeper spacing on the rail vibration is analysed in the vehicle-track vertical coupling model. By comparing the effects of vehicle speed and load on the vibration response of rails under different sleeper spacings, the importance of vehicle system is pointed out. According to the power spectrum density of the vertical rail displacement, the optimal sleeper spacing under the specific line is proposed, and verified via the vibration decay rate. The results show that the sleeper spacing directly affects the first-order Pinned-pinned vibration of the rail, and the effects of the speed and the load on the vibration response are different. In the low-speed section, the impact of the sleeper spacing on the vibration response is smaller, and the larger spacing can be appropriately selected to reduce the number of sleeper to save costs. However, in the high-speed section, the impact is larger, and the speed and the load should be comprehensively considered to select the optimal sleeper spacing.

1 Introduction

With the development of urban rail transit lines in China, the vibration and noise problems caused by the operation of rail trains are also becoming more and more serious. In the track design process, the force of the rail at the sleeper support is directly related to the distance between the sleepers, and the magnitude of the force affects the vibration of the rail. Therefore, the study of sleeper spacing has received extensive attention from scholars [1-4].

In recent years, domestic and foreign scholars have done a lot of research on the issues related to sleeper spacing[2-6]. Wu T X et al. regarded the sleeper spacing and the track bed stiffness as random variables, and analysed their effects on the orbital vibration by numerical simulation. The results show that the dynamic compliance and decay rate of the rail in a certain area are not fixed due to the variation of the sleeper spacing and the stiffness of the track bed [2]. Jin X S et al. used numerical methods to study the influence of sleeper spacing on the initiation and development of tangential orbital wave grinding. In the semi-vehicle track model, the easy-excitation resonance frequency of the wheelset and the track at different speeds is analysed in detail, and the influence of the discrete sleeper support on the normal load of the wheel-track is analysed at different speeds [3]. Batjargal S et al. studied the theory of optimizing sleeper spacing in the Timoshenko beam orbit model. The results show that the optimization of the sleeper spacing is beneficial to reduce the vibration response of the rail in the high frequency band, and the effectiveness of the vibration reduction method is verified by harmonic response analysis [4]. Šestáková J et al. considered that the relevant design and geometric parameters of the track include track alignment, gauge, superelevation and distortion, and evaluated the characteristics of different orbital structures through experiments [5]. Gao L et al. established a finite element model of slab ballastless track vibration to study the factors affecting the Pinned-pinned resonance of the rail. The results show that the change of the sleeper spacing will cause the Pinned-pinned resonance characteristics of the rail to change, which is a sensitive parameter [6]. In summary, most of the above studies are in the track model, and the influence of the vehicle system is rarely considered. Even considering the vehicle, the influence of the sleeper spacing on the rail vibration is studied only under the fixed vehicle speed. Due to the variable speed and passenger capacity of urban rail transit, the different working conditions of vehicles in different sections have great differences, and the operation state of a certain vehicle cannot accurately analyse the role of the sleeper spacing.

In the paper, a vehicle-track coupling model is established. According to the different working conditions of the vehicle system, the influence of the sleeper spacing on the vertical vibration response of the rail is studied, and the importance of vehicle system to rail research is pointed out. The effects of speed and load on the vertical vibration of rails are compared, then the recommendations for the optimal sleeper spacing are proposed for the lines with different characteristics of urban rail transit.

[a] Corresponding author: yp_wen@163.com

2 Vehicle-track coupling Model

Figure 1 shows the vertical dynamics model of the vehicle-track coupled system. The vehicle model is type A and the rail model is UIC60. The vehicle system consists of a carbody, two bogie and four wheelsets, all of which are considered rigid bodies.

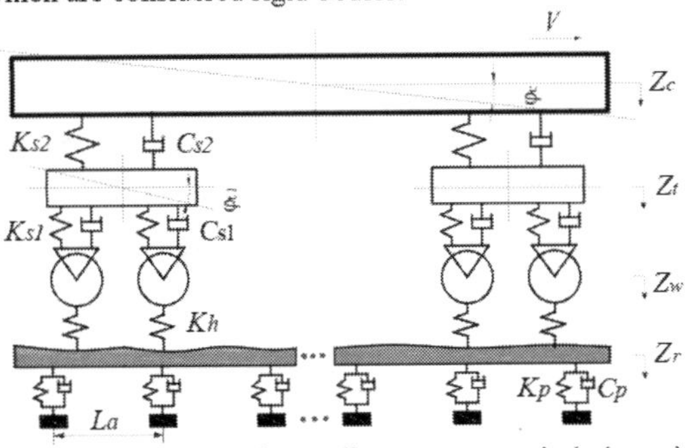

Figure 1. Vehicle-track coupling system vertical dynamics model.

Both the carbody and the bogie have two degrees of freedom of bounce and pitch. Z_c and φ_c are used to indicate the vertical displacement and the pitch angle of the carbody, and Z_t , φ_t are used as the vertical displacement and the pitch angle of the bogie; each wheelset has a degree of freedom in bounce, and Z_w is used to indicate the vertical displacement of the wheelset.

The running speed of the vehicle is V , and the running direction is positive to the right. The track section is a 50 m long pillow-buried ballastless track. The rail is regarded as a simply supported Euler beam, and its vertical displacement is indicated by Z_r . In order to broaden the applicable frequency range of the Euler beam model, the sectional inertia of the rail section is taken as 75% of the value in the table [7]. The remaining parameters are shown in Table 1.

According to the bending dynamics of the beam and the stress of the rail, the vibration differential equation of the rail is established [8]:

$$E_r I_r \frac{\partial^4 Z_r(x,t)}{\partial x^4} + M_r \frac{\partial^2 Z_r(x,t)}{\partial t^2} + \sum_{i=1}^{N_r} \delta(x-x_i) F_{ri}(t) = \sum_{j=1}^{4} \delta(x-x_{wj}) F_{wj}(t) \quad (1)$$

Where, N_r is the total number of fastener nodes on the rail sleeper, that is, the number of sleepers, x_i is the coordinate value of the i -th fastener node on the rail. And F_{ri} is the fastener fulcrum force, and x_{wj} is the coordinate value of the j -axis wheelset. Then F_{wj} is the force of the wheel and rail action.

Table 1. Parameters of vehicle-rail coupling model for a city rail vehicle.

Meaning	Parameter	Value
mass of the carbody	M_c	3.4×10^4kg
mass of the bogie	M_t	2.8×10^3kg
mass of the wheelset	M_w	1.119×10^3kg
mass of per unit length of the rail	M_r	$60.64\text{kg}\cdot\text{m}^{-1}$
moment of inertia for carbody pitch	J_c	$1.686\times10^6\text{kg}\cdot\text{m}^2$
moment of the bogie pitch	J_t	$1.42\times10^3\text{kg}\cdot\text{m}^2$
Elastic modulus of the rail	E_r	$2.059\times10^{11}\text{N}\cdot\text{m}^{-2}$
Sectional inertia of the rail	I_r	$3.217\times10^{-5}\text{m}^4$
vertical equivalent stiffness of the primary suspension	K_{s1}	$3.68\times10^6\text{N}\cdot\text{m}^{-1}$
vertical equivalent stiffness of the secondary suspension	K_{s2}	$1.98\times10^6\text{N}\cdot\text{m}^{-1}$
vertical stiffness of the fastener	K_p	$5\times10^7\text{N}\cdot\text{m}^{-1}$
vertical damping of the primary suspension	C_{S1}	$6.6\times10^4\text{N}\cdot\text{s}\cdot\text{m}^{-1}$
vertical damping of the secondary suspension	C_{s2}	$6.4\times10^4\text{N}\cdot\text{s}\cdot\text{m}^{-1}$
vertical damping of the fastener	C_p	$7.5\times10^4\text{N}\cdot\text{s}\cdot\text{m}^{-1}$
half the length between the truck centers	L_b	7.85m
half the wheelbase	L_w	1.25m
length of the carbody	L_c	21.88m
length of the rail	L_r	50m

Generally, the total length of the rail and the number of sleepers are used to indicate the sleeper spacing:

$$L_a = Lr/Nr \tag{2}$$

Where, L_a is the sleeper spacing.

Combining the vehicle with orbital dynamics equation and perform a Fourier transform on it:

$$\left(-\omega^2 \boldsymbol{M} + j\omega \boldsymbol{C} + \boldsymbol{K}\right)\boldsymbol{Z}(\omega) = \boldsymbol{K}_f \boldsymbol{q}(\omega) \tag{3}$$

Where, $\boldsymbol{M}$, $\boldsymbol{C}$, and $\boldsymbol{K}$ are the mass, damping, and stiffness matrices of the system, $\boldsymbol{Z}(\omega)$ is the output matrix of the system, $\boldsymbol{K}_f$ is the system transformation matrix, and $\boldsymbol{q}(\omega)$ is the input matrix of each wheelset of the vehicle.

The vehicle system and the track system are coupled by wheel-rail contact force, and it is a nonlinear force. In order to facilitate the calculation and make the wheel-rail contact force a linear force, it can be linearized by the wheel-rail contact stiffness, which is specifically expressed as:

$$K_h = \frac{3}{2}\frac{1}{G}P^{\frac{1}{3}} \tag{4}$$

Where, G is the wheel-rail contact constant, and P is the wheel-rail static force.

The wheel-rail static force is the total weight exerted on the rail by the various components of the vehicle system, it can be expressed as:

$$P = 9.8\left(M_c + 2M_t + 4M_w\right) \tag{5}$$

The vehicle system runs on the track at the speed V, and the track irregularity excitation $q_1(\omega)$ is input to the rail, and the displacement Z_i is output, which is the frequency response characteristic function, and it can be expressed as:

$$H(\omega)_{Z_i - q_1} + \omega^2 q_1(\omega)^{-1}\left(-\omega^2 \boldsymbol{M} + j\omega \boldsymbol{C} + \boldsymbol{K}\right)^{-1} \boldsymbol{QK}_f \\ + \omega^2\left[h_1(\omega) \quad \text{L} \quad h_{10+M}(\omega)\right]^T = 0 \tag{6}$$

Where, $h_{11}(w)\text{L}\ h_{10+M}(w)$ is the displacement frequency response function of each mode of the rail.

When the power spectrum density (PSD) of the track irregularity excitation is input, the output rail vertical displacement PSD can be expressed as:

$$G_{Z_i}(\omega) = \left|H(\omega)\right|^2_{Z_i - q_1} G_{q_1}(\omega) \tag{7}$$

Where, $G_{q_1}(\omega)$ is the displacement PSD of the track irregularity excitation.

3 Physical meaning of the sleeper spacing

In the urban rail transit line laying structure, the rail is fixed by the fastener device above the equally spaced discrete sleepers. This form of installation causes the rail to be excited to produce a longitudinal mechanical wave with its standing wave node just at the support of the sleeper, which is the Pinned-pinned vibration of the rail. According to the relevant measured results, the rail radiated noise is the main component of the wheel-rail rolling noise in the frequency range of 500-1500 Hz. The noise peak caused by the first-order Pinned-pinned vibration of the rail is within this frequency range [9]. The first-order Pinned-pinned vibration frequency f_{pp} is related to the rail section type, the rail mass and the sleeper spacing [10], which can be expressed as:

$$f_{pp} = \frac{\pi}{2L_a^{\ 2}}\left(\frac{E_r I_r}{M_r}\right)^{1/2} \tag{8}$$

It can be known from equation (8) that the value of the sleeper spacing directly affects the peak frequency of the Pinned-pinned vibration of the rail. The larger the distance between the sleepers is, the smaller the first-order Pinned-pinned vibration frequency of the rail.

The distance between railway sleepers in China is generally 0.6 m, and the distance between curved sections with a radius less than 800 m is generally 0.625 m. The distance between the sleepers of foreign urban rail transit systems is generally not less than 0.7 m, and the curve section is to ensure the smoothness of the rails from 0.7 m to 0.9 m.

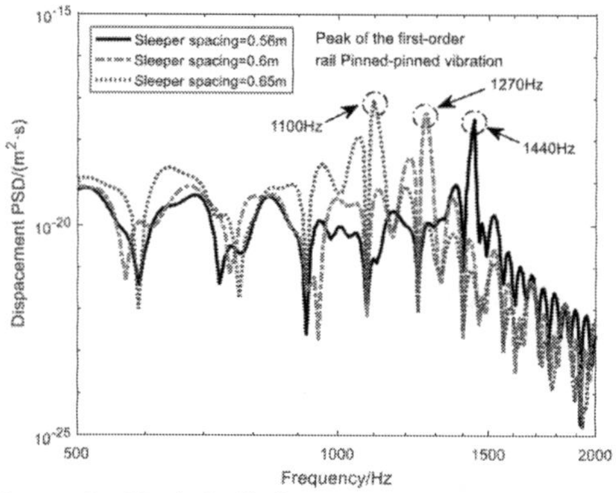

Figure 2. Vertical displacement power spectrum of the midpoint of a 50 m rail.

The paper uses a model of a 50 m long linear orbit. Select a car to move on the track, compare the vertical displacement PSD of the midpoint of the three different sleeper spacings under 80 km/h speed and no-load conditions, as shown in Figure 2. It can be seen from Fig. 2 that in the frequency range of 500-2000 Hz, the vertical vibration response of the rail increases first and then decreases, and the vibration peak is formed at the peak of the first-order Pinned-pinned vibration. When the sleeper spacing is increased from 0.56 m to 0.65 m, the first-order Pinned-pinned vibration peak frequency of the rail is reduced from 1440 Hz to 1100 Hz, and the vibration peak is gradually increased. The change of the value of the sleeper spacing will not only change the first-order Pinned-pinned vibration peak frequency of the rail, but also affect the vertical vibration amplitude of the rail. It is a sensitive parameter affecting the rail vibration in the track structure. Therefore, in the process of laying urban rail transit sleepers, various designs that emphasize the optimal sleeper spacing should be considered.

Figure 2 is an analysis of the sleeper spacing with fixed load and speed. In the following, the effects of speed and load on the vertical vibration of the rail will be further studied in combination with the frequent changes of urban rail transit speed and the large difference in passenger capacity between different lines.

4 Influence of vehicle system on vertical vibration of rail

4.1 Influence of speed

Due to the frequent changes in urban rail transit passenger flow and the length of station spacing, and the frequent start, acceleration and braking of vehicles [11], the running speed of each section is quite different. Under no-load condition, the traversal is performed in the speed range of 0~80 km/h, and the peak of the vertical displacement PSD of the midpoint of the rail is compared with the three kinds of sleeper spacing, as shown in Fig. 3.

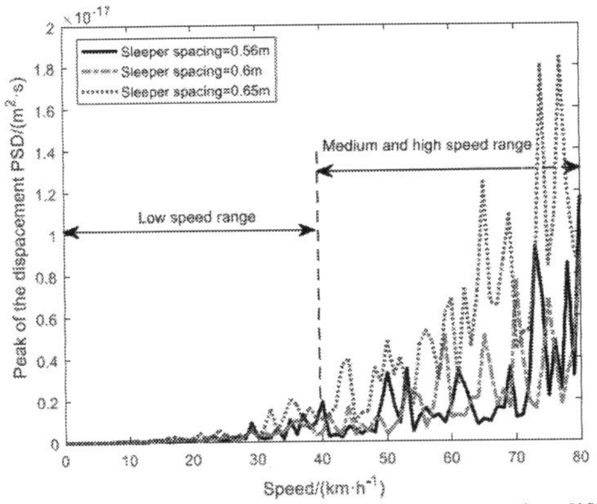

Figure 3. Effect of speed on rail vibration under different sleeper spacing.

Fig. 3 presents that under the three sleeper spacings, the vertical vibration of the rails increases with the increase of the speed, that is, the vibration response of the rails in the high speed range is basically greater than the response in the low speed range. This is because the increase in speed increases the force of action between the wheel and rail, resulting in an increase in rail vibration [12-14]. In addition, it is worth noting that in the low speed range, the change in the value of the sleeper spacing has less effect on the vertical vibration response of the rail during the same speed vehicle operation. In the high-speed range, when the vehicle of the same speed travels on the line with different sleeper spacing, the vertical vibration response of the rail has a big difference. When the sleeper spacing is 0.65 m, the peak curve of the vertical displacement power spectrum of the rail is the highest. Therefore, under different speeds (especially high-speed intervals), choosing the most suitable sleeper spacing is beneficial to control the vertical vibration of the rail and thus control the radiation noise between the wheel and rail.

4.2 Influence of load

The large change in passenger capacity is another significant feature of urban rail transit [11]. Passenger traffic is often large in the downtown area, and vehicles are fully loaded or even overloaded most of the time, while in the suburbs, urban rail transit passenger traffic is generally maintained at medium and below conditions. The difference in passenger flow results in different vehicle load conditions, and the impact of different sleeper spacing lines on the vertical vibration of the rails is also different.

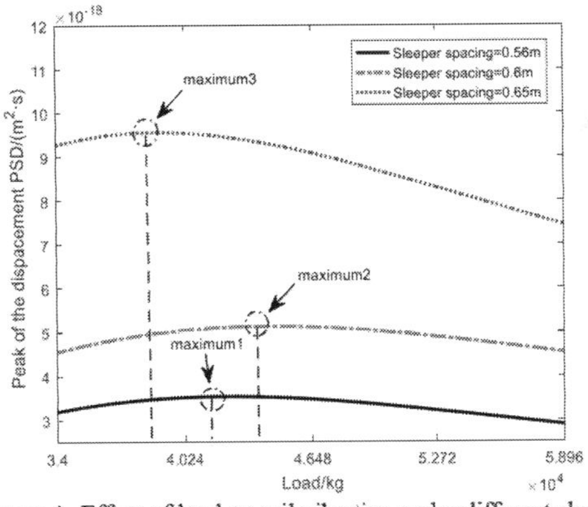

Figure 4. Effect of load on rail vibration under different sleeper spacing.

Figure 4 shows the variation curve of the vertical displacement PSD of the rail from no load to overload during different distances of the sleeper at 80 km/h. It can be seen from Fig. 4 that during the change of the vehicle from no load to overload, the peak value of the vertical displacement PSD of the rail shows a trend of increasing first and then decreasing. However, the maximum curve corresponding to different sleeper spacings appears under different vehicle load values. Moreover, in the process of changing the distance between the sleepers from 0.56 m to 0.65 m, the larger the spacing value, the more severe the vertical response of the rail. That is to say, when the vehicle is running on the line with different sleeper spacing under the same load condition, there is a certain difference in the vibration response of the rail. Therefore, it is necessary to design the most suitable sleeper spacing for a specific track line for different load conditions of the vehicle, so as to achieve the purpose of controlling the vertical vibration of the rail and the radiation noise between the wheel and rail.

https://doi.org/10.1051/matecconf/201929601008

5 Designing of the optimal sleeper spacing

In view of the diversity of vehicle speed and load capacity, the design of the optimal sleeper spacing is studied from the perspective of speed and load.

The paper takes the peak of the vertical displacement power spectral density of the rail as the index to find the optimal sleeper spacing. According to the common sleeper spacing in China, the range of spacing is determined to be between 0.5 m and 0.7 m. In the model, a length of 50 m is used, that is, the number of rails is about 70 to 95. Under the four typical load conditions (no load, heavy load, full load and overload), the traversal selection of the 50 m rail sleeper number in the range of 0~80 km/h is shown in Fig. 5.

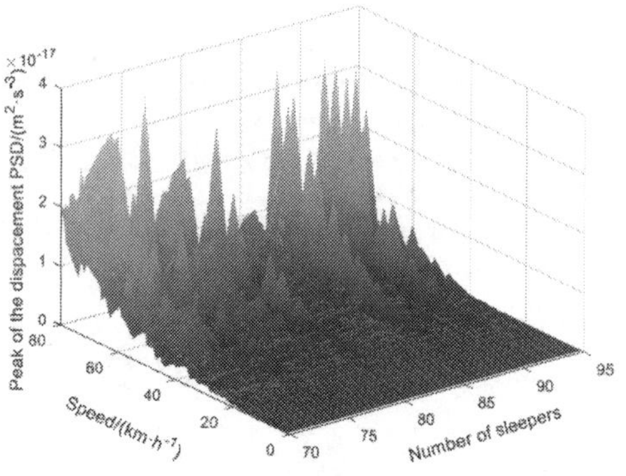

(a). Aw0 working condition.

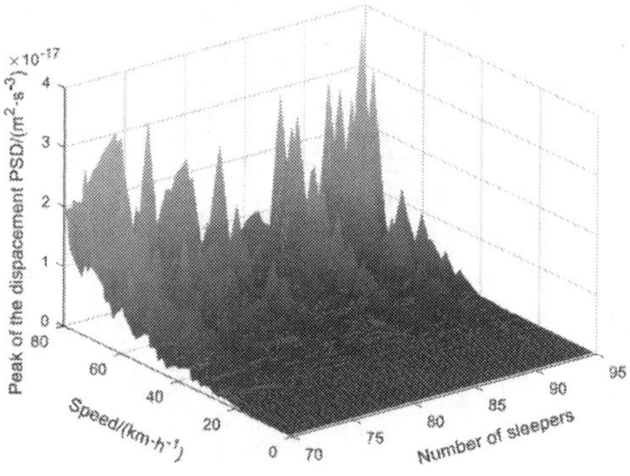

(b). Aw1 working condition.

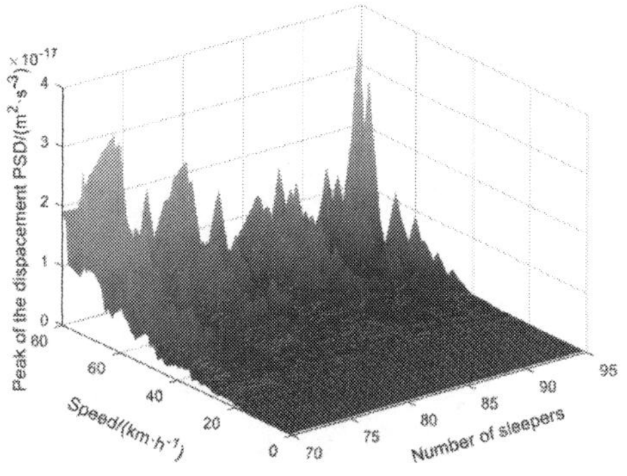

(c). Aw2 working condition.

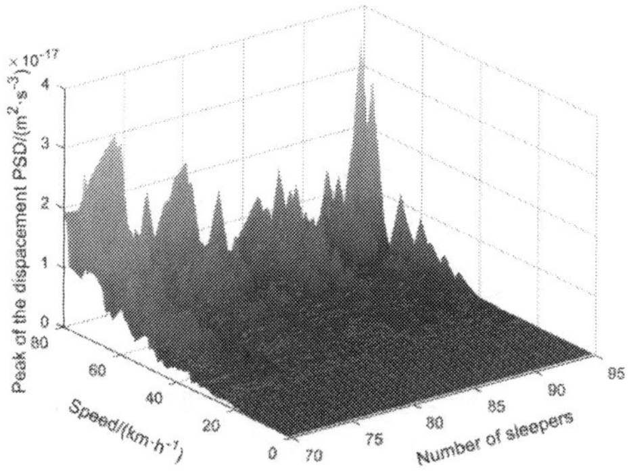

(d). Aw3 working condition.

Figure 5. Vertical vibration response of rail under four load conditions.

Referring to Fig. 5 (a), (b), (c), (d), it can be seen that the rail vertical vibration response is small in the low speed section below 40 km/h regardless of the load condition. Moreover, the influence of the change of the sleeper spacing on the vertical vibration of the rail is also less obvious; In the speed range of 40~80 km/h, the vertical response of the rail gradually increases with the increase of the speed of the same sleeper spacing, which corresponds to Figure 2. In the different load conditions of (a), (b), (c), (d), the magnitude of the ordinate axis corresponding to the number of the same sleeper is also different.

In order to obtain the regularity of the vertical vibration response amplitude of the rail under the four load conditions, the number of sleepers is the abscissa, and the maximum value of the peak of the rail displacement power spectral density corresponding to the number of each sleeper is taken as the ordinate, and four load cases are obtained. The different curves are shown in Figure 6.

It can be seen from Fig. 6 that the maximum value of the peak of the displacement PSD of the rail corresponding to the number of different sleepers forms a curve with multiple peaks and valleys. The valley value corresponds to the optimal sleeper spacing that minimizes the vertical vibration response of the rail under this condition, and the peak value is the opposite. The curves under different load conditions do not overlap completely, but the values on the curve are peak or valley just when the same number of sleepers is used. As indicated in the figure, when the number of sleepers is 75, 83, 87, 89, 91 and 93, that is, when the sleeper spacing is 0.67 m, 0.6 m, 0.57 m, 0.56 m, 0.55 m and 0.54 m, the peak of the vertical displacement PSD of the rail is the smallest, that is, the vertical vibration response is the lowest.

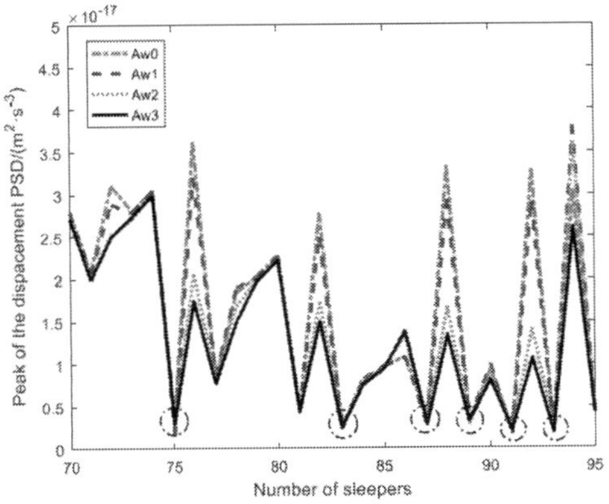

Figure 6. Peak curve of rail displacement spectrum with different number of sleepers.

As can be seen from Fig. 5 and Fig. 6, in the low-speed straight line segment (such as the yard line and the maintenance line), the change of the line sleeper spacing has little effect on the vertical vibration response of the rail regardless of the load capacity. Considering economic factors, it is possible to appropriately select a larger sleeper spacing and reduce the number of laying sleepers to save costs. In the high-speed section (such as the suburban line and the larger station spacing), the change of the sleeper spacing has a greater impact on the vertical vibration response of the rail. Considering the factors of speed and load, combined with the actual topography, it is recommended to select the optimal sleeper spacing at 0.67 m, 0.6 m, 0.57 m, 0.56 m, 0.55 m and 0.54m.

6 Verification

At present, the vibration decay rate is usually used in the study of rail vibration to reflect the ability of vibration to decay along the rail [15-16]. In a frequency range, the greater the vibration decay rate value, the smaller the vibration response of the rail. The paper uses the vibration decay rate to verify the optimal sleeper spacing value.

The vibration decay rate of the rail can be solved by the displacement response, which can be expressed as:

$$r = \frac{20\lg\left|Z_r(X_1)/Z_r(X_2)\right|}{\Delta X} \quad (9)$$

Where, $Z_r(X_1)$ is the vertical displacement of the rail at the excitation point, $Z_r(X_2)$ is the vertical displacement of the rail at the inspection point, and ΔX is the distance between X_1 and X_2.

According to Fig. 6, when the number of sleepers for the 50 m rail is 88, that is, the distance between the sleepers is 0.57 m, and the ordinate is the peak value, indicating that the rail has the largest vertical response when the sleeper is spaced. When the number of sleepers for the 50 m rail is 75, that is, the distance between the sleepers is 0.67 m, the ordinate is the valley value, indicating that the vertical response of the rail is the smallest. Select the two types of sleeper spacing and compare the root mean square value of the vibration decay rate of the rail in the high speed range of 40-80 km/h, as shown in Fig. 7.

It can be seen from Fig. 7 that in the high speed range of 40-80 km/h, the root mean square value of the vibration decay rate of the rail when the sleeper spacing is 0.67 m is significantly larger than the value of the vibration attenuation rate when the sleeper spacing is 0.57 m. That is to say, compared with 88 sleepers, laying 91 sleepers in the 50 m linear rail range makes the rail vibration response smaller, which corresponds to the content of Fig. 6. Therefore, in the vehicle-track coupling model, the speed and load factor of the vehicle system should be fully considered, and the optimal sleeper spacing is selected according to the characteristics of the line, which has obvious effects on reducing the vibration of the rail.

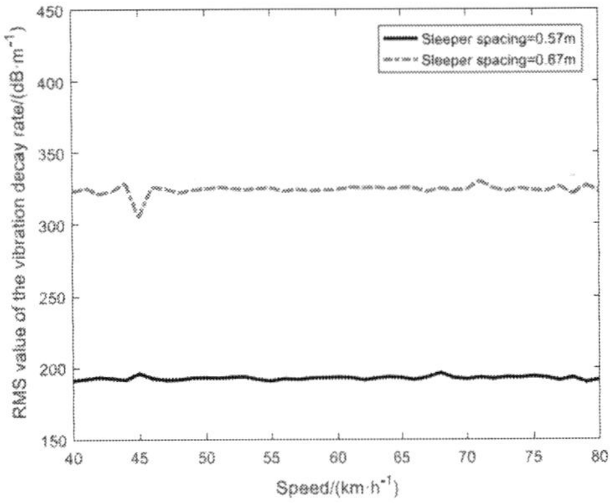

Figure 7. Root mean square value of vibration decay rate of rail under two types of sleeper spacing.

7 Conclusions

The sleeper spacing directly affects the value of the first-order Pinned-pinned vibration peak frequency of the rail. And under different sleeper spacings, the influence of the vehicle speed and the load on the vibration response of the rail is different. Therefore, when designing the sleeper spacing for the line, the role of the vehicle system should be considered comprehensively.

Under different load conditions, the vertical vibration response amplitude of the rail is different when the same sleeper spacing is used. However, the law of vibration response changes with the spacing is basically similar. The extreme values of the vibration response of the rail under different load conditions are consistent with the same number of sleepers.

In the low-speed section, regardless of the load conditions, the change of the sleeper spacing has little effect on the vertical vibration response of the rail. Considering economic factors, it is possible to appropriately select a larger spacing and reduce the number of laying sleepers to save costs. In the high-speed section, the change of the sleeper spacing has a greater influence on the vibration response of the rail. Considering the effects of speed and load on the coupling system, it is recommended to choose the optimum sleeper spacing between 0.67 m, 0.6 m, 0.57 m, 0.56 m, 0.55 m and 0.54 m.

Acknowledgements

This research was partially supported by the National Natural Science Foundation of China (Grant No. 11472176) and the Natural Science Foundation of Shanghai (Grant No. 15ZR1419200).

References

1. Zhang P, Wang A B, Wang Z Q, Xu N and Zhang Z Y 2016 Influence of track parameters on rail Pinned-Pinned vibration. *Urban. Mass. Transit.* **19(12)** 72-76+82
2. Wu T X, Thompson D J 2000 The influence of random sleeper spacing and ballast stiffness on the vibration behaviour of railway track. *Acta. Acust. United. Ac.* **86(2)** 313-321
3. Jin X S, Xiao X B, Wen Z F and Zhong R Z 2008 Effect of sleeper pitch on rail corrugation at a tangent track in vehicle hunting. *Wear.* **265(9)** 1163-1175
4. Batjargal S, Abe K, Koro K 2012 Sleeper spacing optimization for vibration reduction in rails. *J. Comput. Struct. Eng. Inst. Korea.* **25(6)** 569-577
5. Šestáková J, Mečár M 2015 Evaluation of track design and track geometry of the track with unconventional structure of railway superstructure. *Procedia. Eng.* **111** 709-716
6. Gao L, Cui R X, Hou B W and Xiao H 2018 Sensitivity analasis of influence parameters on resonance characteristics of ballasted track rail. *J. Hunan Univ: Nat. Sci. Ed* **45(03)** 115-121
7. Zhao Y, Xiao X B, Han J, Zhao G T and Jin X S 2013 Vertical characteristic and its parameter effect of rail vibration absorber used in high-speed ballasted track. *Chin. J. Mech. Eng.* **49(16)** 17-25
8. Zhai W M, Wang K Y, Cai C B 2009 Fundamentals of vehicle–track coupled dynamics. *Vehicle Syst Dyn.* **47(11)** 1349-1376
9. Thompson D J, Jones C J C 2000 A review of the modelling of wheel/rail noise generation. *J. Sound. Vib.* **231(3)** 519-536
10. Sun F Q, Gu A J, Liu W N 2013 A Study on vibration and transmission characteristics of long solid models under different frequencies. *J. Railway Soc.* **35(2)** 81-86
11. Wen Y P, Li Q, Shang H L and Xu X J 2016 Study on the performance of dynamic vibration absorber for the carbody vibration considering the effects of vehicle-track coupling. *J. Railway Soc.* **35(21)** 53-62
12. Nordborg A 1998 A Parametrically excited rail/wheel vibrations due to track-support irregularities. *Acta. Acust. United. Ac.* **84(5)** 854-859
13. Connolly D P, Kouroussis G, Woodward P K, Alvescosta P, Verlinden O and Forde M C 2014 Field testing and analysis of high speed rail vibrations. *Soil. Dyn. Earthq. Eng.* **67** 102-118
14. Sheng X, Li M, Jones C J C and Thompson D 2007 Using the Fourier-series approach to study interactions between moving wheels and a periodically supported rail. *J. Sound. Vib.* **303(3-5)** 873-894
15. Wu T X 2008 On the railway track dynamics with rail vibration absorber for noise reduction.*J. Sound. Vib.* **309(3)** 739-755
16. Liu H P, Wu T X, Li Z G 2009 Theoretical modeling and effectiveness study of rail vibration absorber for noise control. *J. Sound. Vib.* **323(3)** 594-608

Setting the Configuration Parameters of the Algorithm for the Periodic Vehicle Routing Problem by HPC Power

Ekaterina Grakova[1,a], Jan Martinovič[1], Kateřina Slaninová[1], Kateřina Janurová[1], Vojtěch Cima[1], Martin Golasowski[1], Roberto Montemanni[2], Matteo Salani[3]

[1]*IT4Innovations National Supercomputing Center, VŠB - Technical University of Ostrava, 17. listopadu 15/2172, 708 33 Ostrava, Czech Republic*
[2]*University of Modena and Reggio Emilia, Viale A. Allegri 9, 42121 Reggio Emilia, Italy*
[3]*Dalle Molle Institute for Artificial Intelligence (IDISA, USI/SUPSI), Galleria 2, 6928 Manno, Switzerland*

Abstract. The quality of an optimal solution of the Vehicle Routing Problem is strongly depended on the setting of the configuration parameters of the algorithm. The paper is focused on the introduction of hyperparameter search for solving the Vehicle Routing Problem using a HyperLoom platform for defining and executing scientific pipelines in a distributed environment. To give a concrete example, we focused on Periodic Vehicle Routing Problem for the waste collection. HyperLoom platform was used to define and execute the hyperparameters sweep pipeline. The heuristic algorithm was tested on a real benchmark of the waste collection in Ostrava, Czech Republic. The aim of our case was to effectively combine the minimization of the total travelled distance and the optimization of the fairness of the routes in terms of the standard deviation of a tour length. The waste collection problem was very extensive and computationally demanding, so it was necessary to use high performance computing architecture for testing a large number of different settings of configuration parameters. The experiments were run on the supercomputer Salomon operated by IT4Innovations National Supercomputing Center in the Czech Republic.

1 Introduction

In 2017, the Czech Republic produced 34.5 million tonnes of all waste, of which 1.5 million tonnes of hazardous waste. The Czech Republic is very intensively engaged in waste recycling. From the 34.5 million tonnes of all waste, 84% was recycled, of which 80.5% was material and 3.6% as energy [1]. Every resident in the Czech Republic produces 3,259 kg of all types of the waste per year, of which about 531 kg is the mixed waste. Ostrava is the third largest city in the Czech Republic. Therefore, the problem of the waste collection is very important.

The waste management process is very complex, involving a range of technological methods and disciplines. Collecting problems in transport planning are usually solved by graph theory. One of the types of combinatorial optimization is the Traveling Salesman Problem (TSP), an extension of the TSP problem is Vehicle Routing Problem (VRP) [2], where the problem aims on creating a route of several vehicles with a limited capacity to determine the number of customers.

In this paper, we used the Periodic Vehicle Routing Problem (PVRP). The typical planning period is one day for the basic formulation of VRP. In this case of PVRP, the classical planning is generalized for M days. The difference is that a vehicle is not returning to a depot on the same day (for example, the day it left the depot). In this type of the problem, the goal is to minimize the number of vehicles and time needed to serve the customers (for example, the cost for all rounds). Each customer has a given daily demand, which must be satisfied by one vehicle and by one route. If $M = 1$, we are talking about typical traffic delivery problems. In the case of PVRP, the demand of individual customers can be clustered into groups, creating an individual route for each day. Each customer must be served k-times, where $1 \leq k \leq M$. In the literature, PVRP has received considerable attention. The first works on PVRP were presented by the authors [3], and then there are more papers on this type of the problem [4,5]. Periodic deliveries are used in a lot of applications, including courier services, elevator maintenance and repair, vending machine replenishment, the collection of waste or the delivery of interlibrary loan material.

Heuristic and metaheuristic algorithms are used to solve the problem of waste collection. These algorithms need to set the configuration parameters that improve the quality of the provided solution. In this work, the heuristic algorithm that was used for an exhausted oil collection problem in the area of Bali, Indonesia [6,7] will be used. The formulation types of the problems are very similar. In heuristic algorithm, nine configuration parameters for each type of waste need to be set. Datasets were generated on the basis of cooperation with the collecting company in Ostrava. In total, 28,800

[a] Corresponding author: ekaterina.grakova@vsb.cz

combinations of configuration parameters will be used for algorithm testing.

The aim of this article is to introduce an implementation of a distributed hyperparameter search for the vehicle routing problem solver based on HyperLoom [8]. Using HyperLoom pipeline we are able to sweep through large parameter spaces and discover the optimal solution for the heuristic algorithm. HyperLoom can discover the optimal solution for the heuristic algorithm.

The waste collection problem is very extensive and computationally demanding, so it is necessary to use high performance computing (HPC) architecture for testing a large number of different settings of configuration parameters. The experiments were run on the supercomputer Salomon operated by IT4Innovations.

2 Periodic VRP for waste collection

Betrami and Bodin devised the PVRP problem of waste collection for the city of New York and Washington DC [3]. The first authors represented the concept of PVRP for the daily planning of collection routes. The aim of this concept of Beltrami and Bodin is to minimize the time to be travelled and to select the feasible number of vehicles to be used instead of minimizing the costs of distribution. The solution of the problem was modified by Clarke-Wright algorithm. A Heuristic algorithm is combined here with randomized search procedures [3].

The PVRP for waste collection represented Russel and Igo, who developed heuristic algorithm assigning points to days of the week in order to minimize weekly travel distance (time). The problem presented herein is expanded to more than 750 points [9].

The four-phase approach to the PVRP for the waste collection was presented by Russel and Gribbin in [10], the number of different solution techniques by Chao, Golden, and Wasil in [11]. This article, the aim is to effectively combine the minimization of the total travelled distance and the optimization of the fairness of the routes in terms of the standard deviation of a tour length.

3 Setting the configuration parameters of PVRP algorithm

The authors in the articles [3,12,13] used heuristics or metaheuristics algorithms for the PVRP. The Clark-Wright algorithm is one of the classic heuristics algorithms that is used to solve the VRP and is based on the notion of savings. The solution of the task in Clarke-Wright algorithm is done through the iterations, incremental steps. First of all, the algorithm finds the least effective solution, which is improving each successive step. The solutions carried out in gradual steps monitor and patrol a defined limiting condition. Local Search (LS) is an algorithm, which iteratively translates from solution to solution in a set of acceptable solutions through local changes until we reach the solution, which we consider to be optimal or does not expire the specified search time. Tabu Search (TS) is a metaheuristic algorithm, which uses LS with other rules. LS has tended to end up in a local optimum or in the plains, where several solutions are of equal value. TS, therefore, improves LS by changing the basic rules. No step in the environment would improve the solution, we will use a worsening step (this way we can get from the local optimum), and we will introduce bans (hence the notion of tabu). The genetic algorithm (GA) is the metaheuristic algorithm, which attempts to copy the natural selection process. GA belongs to a broader set of so-called evolutionary algorithms, which mimic techniques from nature such as inheritance, mutation, selection, or crossing.

Optimization algorithms for solving VRP problem involve a large number of design choices and algorithm-specific parameters that need to be set to reach their best performance. The maximum performance of optimization algorithms may depend on the setting of tens to hundreds of parameters. There are a lot of methods to solve the setting of the configuration parameters of the VRP algorithm. Authors in [12] used heuristic search techniques for the setting of the configuration parameters for the metaheuristics algorithms. Another group of parameters tuning of optimization algorithms is focused on statistical modelling approaches [13]. Authors in [12] proposed machine learning approaches.

In machine learning, there exists the term hyperparameters. Hyperparameters present configuration variables that are external to the model and whose values cannot be estimated from data. Hyperparameters tuning can be done using Grid Search, Randomized Search or Bayesian optimization.

4 Test case for PVRP waste collection

In this work, we used mixed waste, paper waste, plastic waste, and glass waste. We prepared a set of 3,000 different locations for the mixed waste, a set of 2,000 different locations for the plastic waste, a set of 1,500 different locations for the paper waste and a set of 1,000 different locations for the glass waste. For each location, we know the expected amount of the waste that accumulates within one week. The travel times between any two locations are known. Every waste collection tour starts at the depot with the empty vehicle. Every vehicle starts from the depot, picks-up the waste from the stops until they are full, dumps them at a dump site. The dump site and the depot for the vehicle are at the same GPS location. Each GPS location is one collection point for the vehicle. The travel time for each route is not allowed to exceed a given bound on working hours corresponding to eight hours. The metric distances were created for each dataset in Ostrava. The situation is displayed in Figure 1.

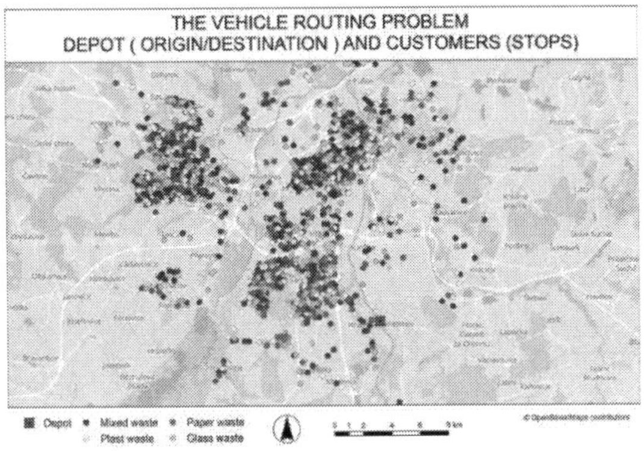

Figure 1. GPS locations in Ostrava.

Planning time for the waste collection was 14 days, including the weekend. The type of collection vehicles depends on the type of the waste.

4.1 HyperLoom for heuristic algorithm

We have explored 7,200 combinations of the configuration parameters for each type of the waste resulting in total of 28,800 unique parameter configurations. We developed a HyperLoom pipeline (see Figure 2) that allows to execute particularly configured instances of the heuristic algorithm in a parallel fashion within a distributed system and dynamically load-balance the workload fairly between the nodes as the instances are being finished.

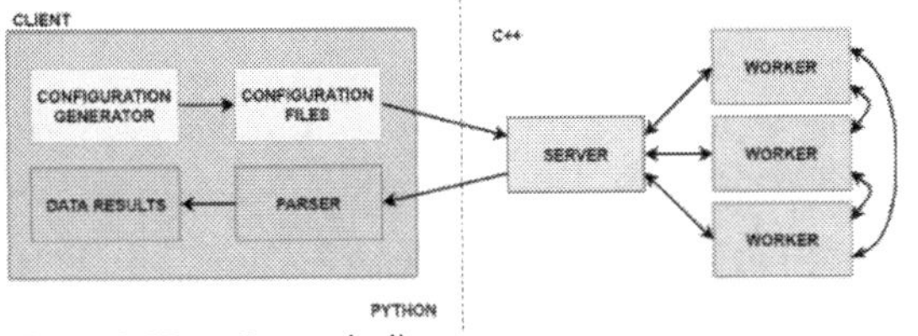

Figure 2. HyperLoom pipeline.

Table 1 contains the settings of configuration parameters for particular types of the waste.

Table 1. Configuration parameters setting for hyperparameter search.

	Mixed waste	Plastic waste	Paper waste	Glass waste
Number of vehicles	50, 51, 52, 53, 54	30, 31, 32, 33, 34	23, 24, 25, 26, 27	23, 24, 25, 26, 27
Depot number	0			
Weight	0.05, 0.50, 0.95			
Random seed	1, 2, 3, 4			
Horizon days	14			
Vehicles capacity	34, 35, 36, 37			
Location per day	100 -109			
Litres per day	200, 250, 300			

The experiments were performed on Salomon supercomputer, where each node has two-twelve-core Intel Xeon processors and 128 GB RAM (2xIntel Xeon E5 – 2680 v3, 2.5 GHz, 12 cores). We have carried out all the experiments on 56 identical physical computational nodes.

Figure 3 illustrates the CPU and memory utilization of a subset of computational nodes during the pipeline execution.

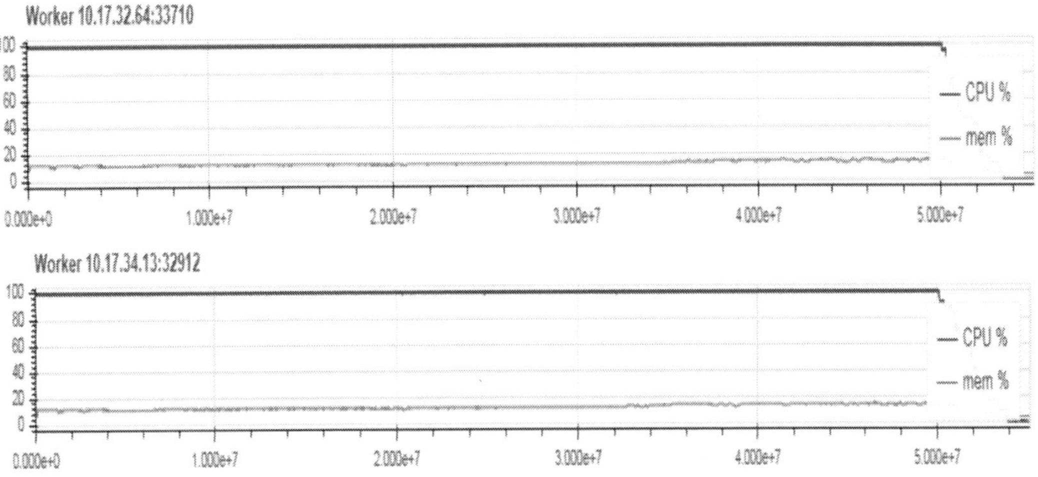

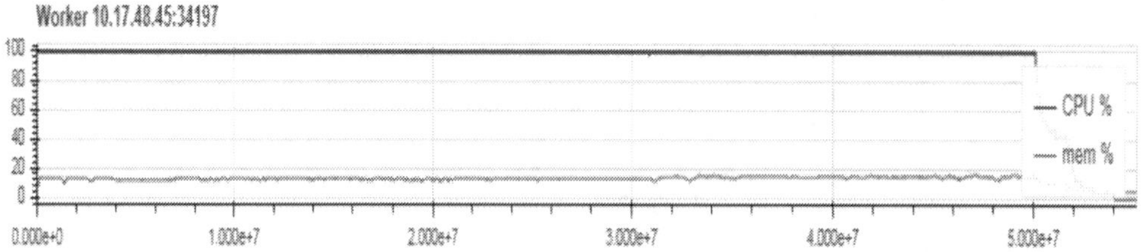

Figure 3. The CPU and memory utilization of computational nodes subset during pipeline execution.

4.2 Test case results

The goal of our work is to optimize the fairness of the routes in terms of the standard deviation of a tour length and to minimize the total of travelled distance. That means that we are looking for a combination of input parameters for which the resulting standard deviation of a tour length value is minimal. If there are more such combinations, we choose one that has a minimum total travelled distance. Figure 4 contains the results of the calculations. The required values of standard deviation vs. total travelled distance are found as far as possible at the bottom left for each type of waste.

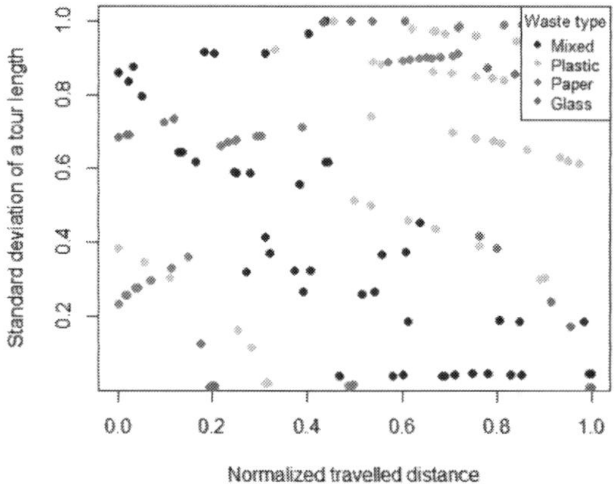

Figure 4. Test case results.

As can be seen from Figure 4, the results for different types of waste show different behaviour. For mixed waste, unlike paper and plastic, there is an increase in the total travelled distance when the standard deviation decreases. This behaviour is also shown by the value of the correlation coefficient $\rho = -0.79$. Plastic and paper waste show the same type of dependency, when the standard deviation increases, the total travelled distance also increases, with the value of correlation coefficients $\rho = 0.43$ and $\rho = 0.59$, respectively. For the glass waste, there is no linear dependence between standard deviation and total travelled distance with the value of correlation parameter $\rho \approx 0$.

The best configuration of input parameters is listed in Table 2. There are listed only those configuration parameters whose values have affected the results because in our case, we deal with a real dataset, the values in Table 2 can allow waste collection companies to reduce their overall waste collection costs.

Table 2. The best configuration parameters setting.

	Mixed waste	Plastic waste	Paper waste	Glass waste
Number of vehicles	53	30	27	27
Weight	0.95	0.95	0.50	0.50
Litres per day	300	250	300	300

5 Conclusion

The goal of our work was to optimize the fairness of the routes in terms of standard deviation of a tour length and to minimize the total travelled distance. We solved this problem by taking the standard deviation as a priority parameter for minimizing. In practice, companies can put more emphasis on minimizing total travelled distance, or on combining both parameters with different weights. In this case, the problem could be solved by the Euclidean distance approach.

Acknowledgements

This work was supported by The Ministry of Education, Youth and Sports from the National Programme of Sustainability (NPS II) project "IT4Innovations excellence in science - LQ1602" and by the IT4Innovations infrastructure which is supported from the Large Infrastructures for Research, Experimental Development and Innovations project "IT4Innovations National Supercomputing Center – LM2015070", and partially by the SGC grant No. SP2019/108 "Extension of HPC platforms for executing scientific pipelines", VŠB - Technical University of Ostrava, Czech Republic.

References

1. Ministry of the Environment of the Czech Republic: https://www.mzp.cz/cz/news/. Accessed 17 Oct. 2017
2. G.B. Dantzig, J. H. Ramser. The Truck Dispatching Problem, Management Science, **6**, 80–9 (1959)
3. E. J. Beltrami, L.D. Bodin. Networks and vehicle routing for municipal waste collection. Networks **4**, chapter 1, 65–94 (1974)

4. S. Coene, A. Arnout, C. Frits Spieksma. The periodic vehicle routing problem. Department of decision sciences and information management (2008)
5. R. Cristian Abreu, J. Elias C. Arroyo. An application of ILS heuristic to periodic vehicle routing problem with heterogeneous fleet and fixed costs. IEEE. 1–9 (2015)
6. D. Weyland, M. Salani, R. Montemanni, L.M. Gambardella. Vehicle routing for exhausted oil collection. Journal of Traffic and Logistics Engineering **1**, chapter 1, 5–8 (2013)
7. E. Grakova, M. Golasowski, R. Montemanni, K. Slaninová, J. Martinovič, J. Jamal, K. Janurová, M. Salani. Hyperparameter Search in Periodic Vehicle Routing Problem. Int. Conf. on Intelligent Traffic and Transportation (ICITT 2018) (2019)
8. V. Cima, S. Böhm, J. Martinovič, J. Dvorský, K. Janurová, T. V. Aa, ... & V. Chupakhin. HyperLoom: a platform for defining and executing scientific pipelines in distributed environments. In Proceedings of the 9th Workshop and 7th Workshop on Parallel Programming and RunTime Management Techniques for Manycore Architectures and Design Tools and Architectures for Multicore Embedded Computing Platforms. 1-6 (2018)
9. R. A. Russell, W. Igo. An assignment routing problem, 2 (1979)
10. R. A. Russell, D. Gribbin. A multiphase approach to the period routing problem (1991)
11. I. - M. Chao, B. L. Golden, E. Wasil. An improved heuristics for the period vehicle routing problem. (1995)
12. P. L. N. U. Cooray, T. D. Rupasinghe. Machine Learning-Based Parameter Tuned Genetic Algorithm for Energy Minimizing Vehicle Routing Problem. Journal of Industrial Engineering, 13 (2017)
13. M. Mirabi, N. Shokri, A. Sadeghieh. Modeling and Solving the Multi-depot Vehicle Routing Problem with Time Window by Considering the Flexible end Depot in Each Route. **3**, 1373-1390 (2016)

Study on the Vibration Suppression Method of Urban Railway Vehicles Based on a Composite Dynamic Vibration Absorber

Qian Sun[1], Yongpeng Wen[1, a] and Yu Zou[1]

1 School of Urban Railway Transportation, Shanghai University of Engineering Science, 333 Long Teng Road, Shanghai 201620, China

Abstract. To reduce the bounce and the pitch vibration of carbody, a vertical dynamic model for urban rail vehicles is established to analyze the vibration response of the carbody in the low frequency range. In this paper, different methods of single-degree-of-freedom dynamic vibration absorber to suppress the vibration for carbody are investigated. The limits of single-degree-of-freedom dynamic vibration absorber to the vibration reduction effect of carbody are pointed out. After that, the design of a composite dynamic vibration absorber including a double oscillator structure is introduced. A vibration discreteness index is used to evaluate dynamic vibration absorbers with various designs for the vibration damping performance. Finally, the vibration reduction performance of the composite dynamic vibration absorber is verified by Sperling's riding index. The results demonstrate that the performance of the single degree of freedom dynamic vibration absorber attached to a carbody may increase the vibration within a partial scope, when the peak frequency of vibration is far away from the design frequency. The installation of the composite dynamic vibration absorber vibration provides gentler running experience for passengers.

1 Introduction

During the operation of rail transit vehicles, the vibration caused by rail irregularities is transmitted to the carbody through the suspension system, which has a certain impact on the ride stability and comfort of passengers. On the one hand, the vibration energy of the carbody is weakened by the transmission and diffusion of elastic waves; on the other, the damping effect depending on the vibration medium is weakened. The mass-spring-damping structure of the dynamic vibration absorber (DVA) has become a representative device for effectively suppressing vibration [1-3].

To effectively improve the stability of vehicle operation, domestic and foreign scholars have conducted a lot of research on the DVA to suppress the vertical vibration reduction of the carbody in several years [4-16]. Foo E. et al. adopted the principle of "skyhook damping" to suppress the elastic vibration of the carbody due to the lightweight design by installing a DVA on the carbody [4]. Wu P B. et al. proposed to install the under-vehicle equipment on the undercarriage as a DVA to suppress the elastic vibration of the EMU carbody, and analyzed the influence of the suspension parameters of the equipment on the vibration of the carbody. Theoretical research found that light equipment and heavy equipment are installed away from the centre of the carbody, the vibration damping performance of the DVA is limited [5-8]. Based on the theory of DVA, Wen Y P. et al proposed a design method of a single-degree-of-freedom (SDOF) DVA, which can effectively reduce the bounce vibration of the carbody, and comprehensively consider the vibration characteristics at different working conditions [9-11]. Based on the fixed-point theory, the frequency ratio and damping ratio of the DVA with negative stiffness solved by Shen Y et al. and the simulation results that the DVA can greatly reduce the resonance amplitude, widen the vibration frequency range, and even keep the amplitude of the main system small in the whole frequency range [12-13]. Combing with multibody dynamics software Simpack, Tomioka T et al. put forward a continuous model in which the carbody seen as elastic which may realize the three-dimensional elastic vibration control of the carbody by the DVA [14-16]. To sum up, most of the previous DVAs are used to control the elastic vibration of the carbody, and few of them are used to control the rigid vibration, especially the dynamic DVAs which can absorb both the bounce and pitch vibration of the carbody. Thus, there is lack of a damping method for simultaneously suppressing the nodding and floating vibration in the rigid vibration of rail vehicles.

In this paper, combined with vibration characteristics of urban rail vehicles to establish the vertical dynamic model of the elastic carbody of urban rail including the composite vibration absorber (CDVA), and the double oscillator structure of the CDVA with two design frequencies to suppress the bounce and pitch vibration of the carbody is confirmed, and puts forward numerical indexes for quantitatively evaluating the vibration damping performance of the carbody. Finally, the paper compares and analyses the design of the SDOF DVA.

[a] Corresponding author: yp_wen@163.com

https://doi.org/10.1051/matecconf/201929601010

The study is conducted from the perspective of the comparison between SDOF DVA and CDVA.

2 Establishment of system model

Fig. 1 is a vertically coupled vibration model of a vehicle system including a composite vibration absorber.

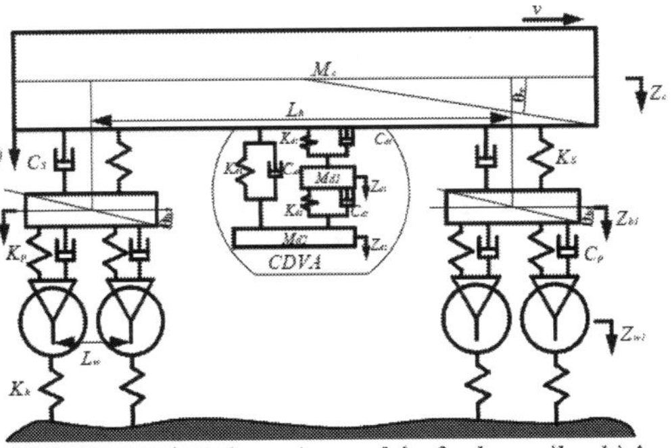

Figure 1. The vertical dynamics model of urban rail vehicles with CDVA

The eight components in the vehicle system include 12 degrees of freedom, respectively corresponding to the two degrees of freedom of the bounce motion Z_c and the pitch motion θ_c of the carbody, the four degrees of freedom of the bounce motion Z_{bi} $(i=1,2)$ of the bogie and the pitch motion θ_{bi} $(i=1,2)$ at the I and II position end, the four degrees of freedom of the bounce motion Z_{wi} of the two front and rear wheelsets and the two degrees of freedom of the bounce motion Z_{di} $(i=1,2)$ of the CDVA (including two oscillators and vibration coupling between oscillator 1 and 2), and the carbody, the bogies and the wheelsets are regarded as rigid bodies. The velocity of the vehicle is v, t is a time variable. The parameters of the model are shown in Table 1.

2.1 Solution of model

According to Lagrange equation, the expression of vertical dynamics of multi-free vehicle system including CDVA can be formulated as below:

$$\begin{bmatrix} \boldsymbol{M}_{cc} & \mathbf{0} \\ \mathbf{0} & \boldsymbol{M}_{dd} \end{bmatrix}\begin{Bmatrix} \ddot{\boldsymbol{Z}}_c \\ \ddot{\boldsymbol{Z}}_d \end{Bmatrix}+\begin{bmatrix} \boldsymbol{C}_{cc} & \boldsymbol{C}_{cd} \\ \boldsymbol{C}_{dc} & \boldsymbol{C}_{dd} \end{bmatrix}\begin{Bmatrix} \dot{\boldsymbol{Z}}_{cc} \\ \dot{\boldsymbol{Z}}_{dd} \end{Bmatrix} + \begin{bmatrix} \boldsymbol{K}_{cc} & \boldsymbol{K}_{cd} \\ \boldsymbol{K}_{dc} & \boldsymbol{K}_{dd} \end{bmatrix}\begin{Bmatrix} \boldsymbol{Z}_c \\ \boldsymbol{Z}_d \end{Bmatrix}=\begin{Bmatrix} \boldsymbol{F}_c \\ \boldsymbol{F}_d \end{Bmatrix} \quad (1)$$

Where, the footnotes c and d respectively represents the vehicle system without the composite vibration absorber and with the CDVA; **M** , **C** , **K** and **F** respectively represents the mass matrix, damping matrix, and **Z** are the generalized acceleration vector, the generalized velocity vector and the generalized displacement vector of components:

$\dot{\boldsymbol{Z}}$ and $\boldsymbol{Z}$ are the generalized acceleration vector, the generalized velocity vector and the generalized displacement vector of components.

Considering the wheelset-time lag characteristics between the four wheelsets in vehicle operation, the system is transformed from multi-excitation input into single, thus obtain the relation matrix between the irregular excitation of the other wheelsets and the first wheelset:

$$\boldsymbol{q}(\omega)=\boldsymbol{Q}q_1(\omega) \quad (2)$$

Where, **Q** is the time delay relation matrix between the first wheel and other wheelsets, and the specific expression is:

$$\mathbf{Q}=\begin{bmatrix} 1 & e^{-j\omega T_1} & e^{-j\omega T_2} & e^{-j\omega T_3} \end{bmatrix}^T \quad (3)$$

Where u is the velocity of the vehicle, $T_1=2L_w/u$, $T_2=2L_b/u$, $T_3=2(L_w+L_b)/u$.

Taking the track irregularity excitation q_1 as the input, the equation (1) can be presented as:

Table 1. Urban rail vehicles dynamic parameters and meanings.

Meaning	Parameter	Value
carbody mass	M_c	39t
wheelset mass	M_w	1.1185t
bogie mass	M_b	2.6t
length of carbody	L	21.88m
half-length between truck centers	L_b	7.85m
half wheelbase	L_w	1.25m
vertical equivalent stiffness of primary suspension	k_p	2400kN·m^{-1}
vertical equivalent stiffness of secondary suspension	k_s	700kN·m^{-1}
vertical damping of primary suspension	c_p	50kN·s·m^{-1}
vertical damping of secondary suspension	c_s	130kN·s·m^{-1}
moment of inertia for carbody nodding	I_c	2300t·m^2
moment of inertia for bogie nodding	I_b	1.424t·m^2

$$\left(-\omega^2\mathbf{M}+j\omega\mathbf{C}+\mathbf{K}\right)\mathbf{Z}(\omega)=\mathbf{K}_f\mathbf{q}_1(\omega) \tag{4}$$

Taking the influence of composite vibration absorber into consideration, **M** is mass matrix of components in the vehicle system, **C** and **K** are damping and stiffness matrix of system considering influence of composite vibration absorber respectively, $\mathbf{K}_f$ transfer matrix between vehicle and wheel excitation.

To analyze the vibration characteristics of rail vehicle system, the frequency response function $H(\omega)_{\ddot{Z}_i-q_1}$ of acceleration response of each component of the vehicle system to random irregularities q_1 is solved:

$$H(\omega)_{\ddot{Z}_i-q_1}=-\omega^2 q_1(\omega)^{-1}Z_i(\omega)=-\omega^2 h_i(\omega) \tag{5}$$

Where, $q_1(\omega)$ is acceleration power spectral density (PSD), $h_i(\omega)(i=1,2,\cdots,12)$ the displacement vector is the system frequency response characteristic. $h_1(\omega)$ is the displacement frequency response characteristics of bounce motion of carbody, $h_8(\omega)$ is displacement frequency response characteristics of pitch motion of carbody; $h_{11}(\omega)$ and $h_{12}(\omega)$ are displacement frequency response characteristics of two oscillator of CDVA, thus the vibration response at any position a of the carbody is:

$$H(a,\omega)=H_1(\omega)+(0.5l-a)H_8(\omega) \tag{6}$$

Where, $H_1(\omega)$ is the frequency response function corresponding to the displacement frequency response $h_1(\omega)$ of the bounce motion of the carbody; $H_8(\omega)$ is the frequency response function corresponding to the displacement frequency response $h_8(\omega)$ of the pitch motion of the carbody.

$$G_{\ddot{Z}_i}(\omega)=\left|H(a,\omega)\right|^2_{\ddot{Z}_i-q_1}G_{q_1}(\omega) \tag{7}$$

Where, $G_{q_1}(\omega)$ is track irregularity excitation.

2.2 The natural frequency of CDVA

The vibration equations of the CDVA with upper and lower oscillators designed is as follows:

$$\begin{cases} M_{d1}\ddot{z}_{d1}=-k_{d1}z_{d1}-c_{d1}\dot{z}_{d1}- \\ k_{d2}\left(z_{d1}-z_{d2}\right)-c_{d2}\left(\dot{z}_{d1}-\dot{z}_{d2}\right) \\ M_{d2}\ddot{z}_{d2}=-k_{d3}z_{d2}-c_{d3}\dot{z}_{d2}+ \\ k_{d2}\left(z_{d1}-z_{d2}\right)+c_{d2}\left(\dot{z}_{d1}-\dot{z}_{d2}\right) \end{cases} \tag{8}$$

Where, M_{d1} and M_{d2} are the masses of the upper and lower oscillators respectively; K_{d1} and C_{d1} (K_{d3} and C_{d3}) are stiffness and damping between the upper oscillator 1 (lower oscillator 1)and the carbody respectively; Z_{d1} and Z_{d2} are the vertical displacements of oscillator 1 and oscillator 2 in the two-degree-of-freedom model.

When the damping effect is not considered, the natural frequency of the CDVA can be written as:

$$\begin{vmatrix} K_{d1}+K_{d2}-M_{d1}\omega^2 & -K_{d2} \\ -K_{d2} & K_{d3}+K_{d2}-M_{d2}\omega^2 \end{vmatrix}=0 \tag{9}$$

Where, $\omega=2\pi f_d$, f_d is the natural frequency of vibration absorber.

According to equation (9), obviously, natural frequency of the CDVA is determined by the mass and stiffness of the two oscillators, and the magnitude of the natural frequency affects the damping effect of the absorber. By determining the mass of the upper and lower oscillators, also, K_{d2} equals K_{d3} , the relationship between natural frequency and the mass or stiffness of the absorber can be obtained by simplifying:

$$f_{d1,2}{}^2=\left(AK_{d1}+BK_{d2}\right)\mp \sqrt{\left(AK_{d1}-CK_{d2}\right)^2+DK_{d1}K_{d2}} \tag{10}$$

Where, A , B , C and D all can be expressed by M_{d1} and M_{d2} .

3 Vibration response of the carbody

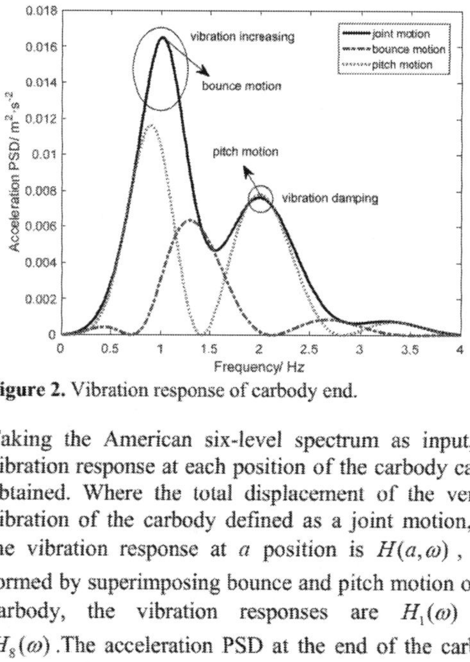

Figure 2. Vibration response of carbody end.

Taking the American six-level spectrum as input, the vibration response at each position of the carbody can be obtained. Where the total displacement of the vertical vibration of the carbody defined as a joint motion, and the vibration response at a position is $H(a,\omega)$, it is formed by superimposing bounce and pitch motion of the carbody, the vibration responses are $H_1(\omega)$ and $H_8(\omega)$ .The acceleration PSD at the end of the carbody when the velocity is 80 km/h and the vehicle is unloaded is selected as an example for analysis, as shown in Fig. 2:

As can be seen from Fig. 2, the vibration response at the end of the carbody is observed with the acceleration PSD as the ordinate. in the low frequency band, the carbody closing motion produces two obvious peaks, the

https://doi.org/10.1051/matecconf/201929601010

first vibration peak is dominated by bounce vibration, and the second vibration peak is dominated by carbody pitch vibration. The vibration response of the carbody closing motion is not a simple linear superposition of the bounce motion and the pitch motion. There is also a coupling vibration between the two motions. Thus, the vibration shows different degrees of enhancement or attenuation on the acceleration PSD.

4 Disadvantages of the SDOF DVA

In the rail vehicle system, the carbody is regarded as the main vibration system. Based on the resonance principle of the DVA, the SDOF DVA can absorb the vibration energy of the carbody. As additional equipment, the quality of the vibration absorber ensures its vibration reduction effect, and the restrictions of factors such as economy and difficulty of arrangement should be considered. Following research, the mass ratio of the vibration absorber is selected as: μ =0.1.

On the premise of determining the quality of the absorber, the key to vibration reduction of a SDOF DVA is to determine the design frequency f_d . When the vehicle is operating at 80 km/h with no load, the carbody has two different vibration peak frequencies. The peak frequency of body heave motion is 1.01 Hz and the peak frequency of pitch motion is 2.00 Hz respectively. For the above frequencies, when designing a SDOF DVA, there are two designs I and II are designs of SDOF DVA with different design frequencies, and the vibration reduction effect is shown in Fig. 3.

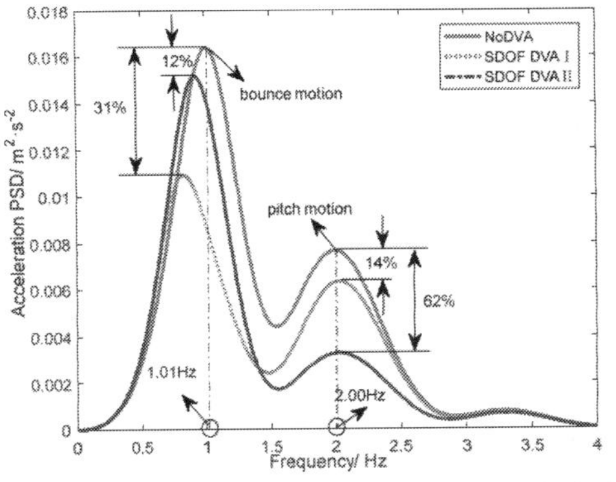

Figure 3. Vibration reduction performance of single-degree freedom vibration absorber

As can be seen from Fig. 3, the SDOF DVA designed for different vibration peaks can effectively reduce the bounce vibration of the carbody, especially when the SDOF DVA is designed for 1.01 Hz, its vibration reduction effect on the bounce motion of the carbody at the peak frequency is about 31%; When the design of SDOF DVA is designed for 2.0 Hz, the damping effect of the SDOF DVA on the pitch motion of the carbody at the peak frequency is 62%; However, the effect of SDOF DVA design on the peak frequency of pitch motion and that of SDOF DVA design on the peak frequency of bounce vibration are only 14% and 12%, respectively. Because the SDOF DVA can only be designed for a certain target vibration reduction frequency, also the target vibration reduction frequency is single.

To sum up, the SDOF DVA can achieve the effect of wide-band vibration reduction to a certain extent. However, when there are two target vibration reduction frequencies, and the design frequency of the DVA is far away from a certain peak frequency of vibration, the design of the SDOF DVA will have limitations and the vibration reduction effect is limited. Therefore, the paper proposes a method of vibration reduction of the carbody by using the CDVA.

5. Superiority of Compound DVA

According to the design of the SDOF DVA, when the vehicle at a velocity of 80 km/h under no-load conditions, the bounce and pitch vibration of the carbody are dominant in the frequency range of 0~4 Hz, and there are two different peak frequencies on 1.01 Hz and 2.00 Hz, the target vibration damping frequency f_{j1} and f_{j2}, they are correspond to the natural frequencies of the CDVA. With equation (9), thus the natural the frequency of the CDVA is determined by the mass and stiffness of the two oscillators. On the premise that the total mass of the absorber is constant, the mass and stiffness of the oscillators need to be determined, so, the stiffness can be expressed by the mass and natural frequency of the absorber:

$$\begin{cases} K_{d1} = \dfrac{\left[4\lambda - \left(2M_{d1}M_{d2} + 4M_{d2}{}^{2}\right)\cdot(\lambda-\tau)\right]}{4M_{d1}M_{d2}{}^{2} + M_{d2}{}^{3}} \\ K_{d2} = \dfrac{4(\lambda-\tau)}{4M_{d1} + M_{d2}} \\ K_{d3} = \dfrac{4(\lambda-\tau)}{4M_{d1} + M_{d2}} \end{cases} \tag{11}$$

Where, λ and τ can be represented by , M_{d1} and M_{d2} f_{d1} and f_{d2} as follows:

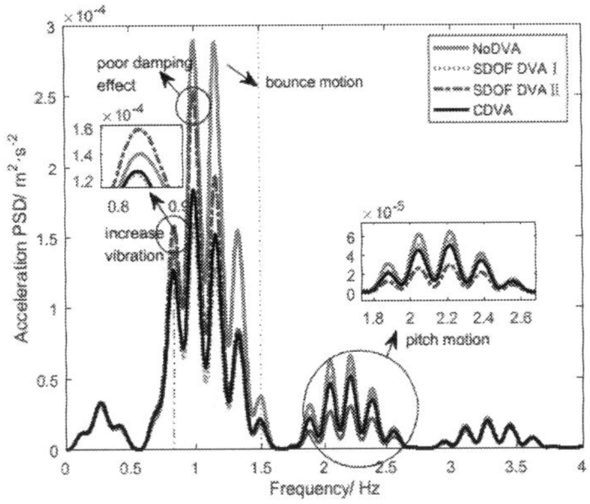

(a) 10 km/h

https://doi.org/10.1051/matecconf/201929601010

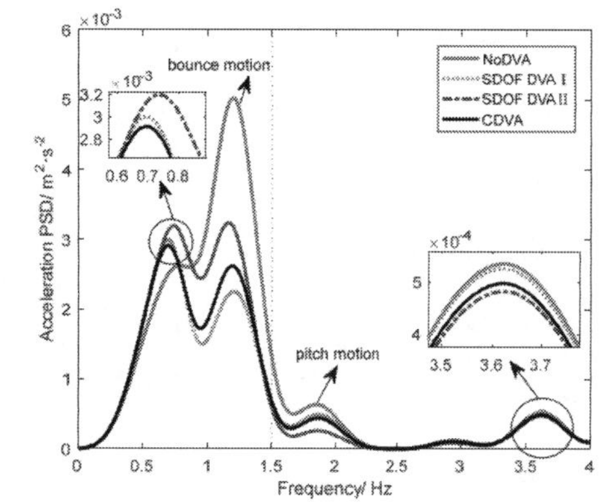

(b) 45 km/h

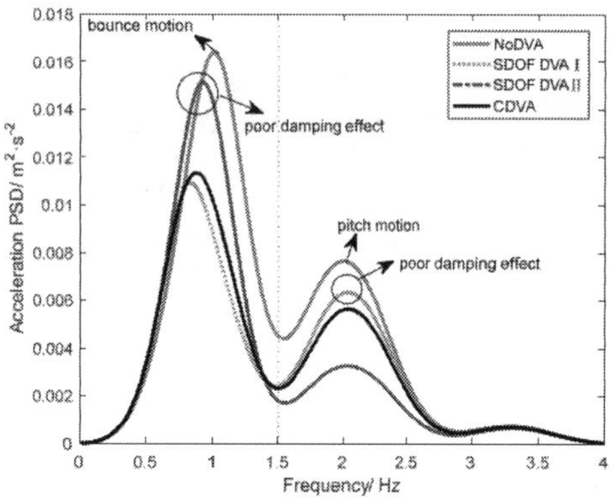

(c) 80 km/h

Figure 4. Influence of SDOF DVA and CDVA on vertical vibration of rail carbody

$$\lambda = M_{d1}M_{d2}\pi^2(f_{d1}^{\ 2} + f_{d2}^{\ 2}) \tag{12}$$

$$\tau = M_{d2}\pi^2\sqrt{M_{d1}^{\ 2}\left(f_{d2}^{\ 2} - f_{d1}^{\ 2}\right) - M_{d1}M_{d2}f_{d1}^{\ 2}f_{d2}^{\ 2}} \tag{13}$$

When the natural frequencies of the absorber are determined, the structural parameters of the CDVA can be obtained.

In order to verify the damping effect of CDVA, considering the frequent changes in vehicle velocity caused by the short distance between urban vehicle traffic stations and vehicle starting and braking frequently, the acceleration PSD at the end position of the carbody under three vehicle velocity conditions, at low velocity 10 km/h, at medium velocity 45 km/h and at high velocity 80 km/h, is investigated, the damping effect of CDVA is compared with the SDOF DVA, as shown in Fig. 4.

As can be seen from Fig. 4, with the change of the vehicle velocity of the urban rail vehicle, the difference reflected on the vibration response at the end of the carbody lies in difference for the vibration characteristics. The CDVA is used to reduce the vibration of the carbody at different operating velocities of the carbody. Although the vibration reduction effect of the CDVA is insufficient compared with the design of the SDOF DVA for the single peak frequency, the vibration reduction effect of the CDVA is obvious at the peak frequencies of floating, sinking and nodding movements, and there is no obvious vibration increase phenomenon in the design of the SDOF DVA in Fig. 4(b). Therefore, the CDVA is not confined to vibration reduction aiming at a certain peak value. By dispersing the total mass of the DVA onto the two oscillators and utilizing the targeted vibration reduction effect of the respective design frequencies of the two oscillators, the problem of the design limitation of the SDOF DVA is treated in a compromise way to realize the effect of wider frequency vibration reduction of the carbody. Compared with the Fig.5, acceleration PSD with the increase of vehicle velocity increases in order of magnitude, and the peak values in each frequency band tend to be concentrated from scattered. This is because when urban rail vehicles are running at high velocity, the unfavourable wavelength that has the greatest impact on the carbody vibration gradually increases, and the irregularity of the track in this wavelength range will excite the body with lower natural vibration frequency. The shorter the wavelength, the more times the internal excitation occurs between units.

To describe the vibration reduction effect of the CDVA and the SDOF DVA in the bounce vibration and pitch vibration frequency bands more intuitively, the vibration is divided into two frequency bands: the first band is 0~1.5 Hz, mainly bounce vibration; The second segment is 1.5~4 Hz, mainly pitch vibration. The vibration discreteness in each frequency band is calculated separately. Considering the vibration characteristics of the vehicle during operation, the bounce vibration of the carbody is significantly larger than the pitch vibration in numerical value. Therefore, when analyse the vibration damping effect of the CDVA and the SDOF DVA quantitatively, the vibration discreteness of each frequency segment obtained needs to be multiplied by the weighting coefficient of the corresponding frequency segment, namely, the comprehensive index of vibration discreteness. The value is small, which means the vibration energy under this velocity condition is small. The specific expression is:

$$CVD = \sum_{k=1}^{2}\left(\sum\left(G_{\ddot{Z}c}\left(f_{jk}\right) - \overline{G_{\ddot{Z}c}\left(f_{jk}\right)}\right)^2 \cdot a_k\right) \tag{14}$$

Where, a_k ($\sum_{k=1}^{2} a_k = 1$) is weighted coefficient of different vibration frequency segments, $G_{\ddot{Z}c}\left(f_{jk}\right)$ is acceleration PSD of carbody vibration acceleration at f_j Hz in k th vibration frequency band. $\overline{G_{\ddot{Z}c}\left(f_{jk}\right)}$ is the average of the acceleration PSD of carbody vibration in the range of k th frequency.

As can be seen from Fig. 5, the CDVA has the lower index of vibration dispersion under the conditions of low velocity of 10 km/h, medium velocity of 45 km/h or high velocity of 80 km/h, that is to say, the lower the vibration dispersion degree of the CDVA is compared with the SDOF DVA, and the obvious the vibration reduction effect is CDVA, SDOF DVA I is after, the SDOF DVA II is last. Because of the double-oscillator structure of the CDVA, the distribution of mass and the targeted damping of the two design frequencies contribute significantly to the damping at the peak frequency of the carbody. As can be seen from Figs. 5(a), (b) and (c) that with the increase of velocity, the dispersion index of the vehicle increases gradually in numerical, that is, the greater the dispersion degree of the acceleration PSD value of the carbody in the modified frequency band, the greater the vibration response of the vehicle. Own to the increase of velocity, the intensification of the function between the wheel and rail of the vehicle, thus the vibration energy of the carbody increases, and the greater the acceleration PSD value, the greater the dispersion degree.

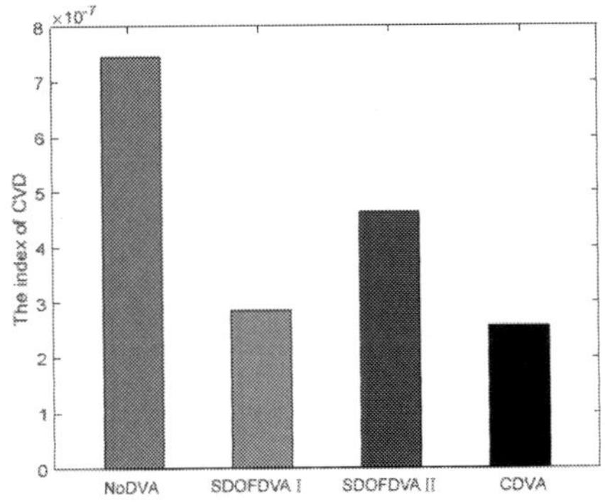

(a) 10 km/h

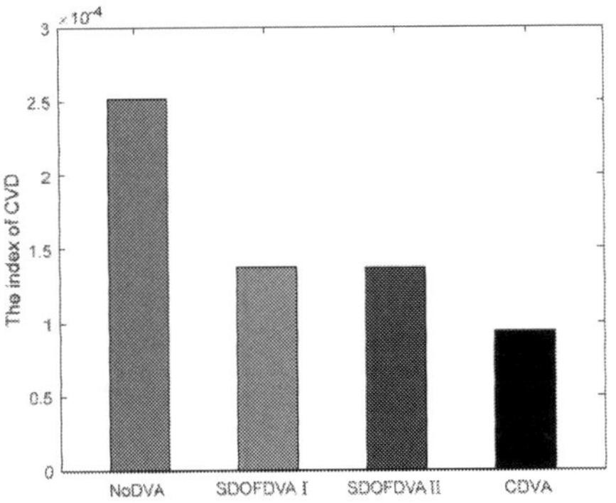

(b) 45 km/h

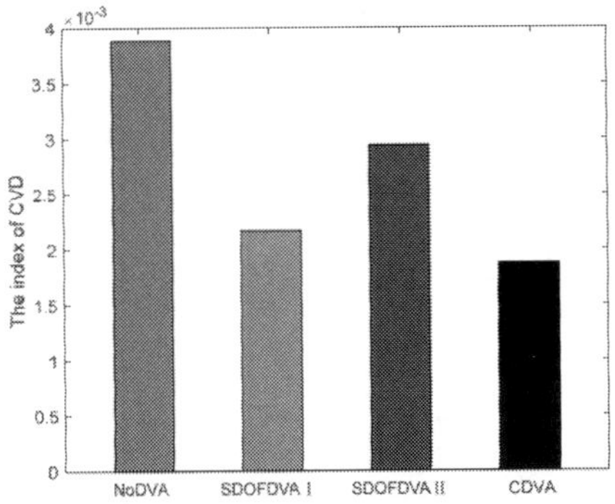

(c) 80 km/h

Figure 5. Influence of SDOF DVA and CDVA on vertical vibration of rail carbody.

6. Sperling's riding index verification

Ride comfort depends on different dynamic performance standards and subjective feelings of passengers. Vibration of different factors such as vehicle conditions, track areas and operating conditions will reduce ride comfort of passengers. According to the specific needs of the railway management department for traffic details, there are also a wide range of methods for evaluating ride comfort [17-18]. Sperling's riding index is adopted in this paper. The result is intuitive and more suitable for comparison of two or more operating conditions.

In this paper, the product of impulse and vibration kinetic energy is used as a measurement standard to evaluate the running stability of vehicles. The formula is:

$$W_z = 2.7^{10}\left(z_0 f^5 F(f)\right)^{0.5} = 0.896^{10}\left(F(f)/f\right)^{0.5} \quad (15)$$

Where, z_0 is amplitude, f is frequency, $a = z_0(2\pi f)^2$ is acceleration, $F(f)$ is weighted coefficient related to vibration frequency

For vertical vibration, there are four different expressions of the weighting coefficient for the difference of the value range. Through spectrum analysis, the stationarity index of each frequency band is solved. Therefore, the total stationarity index W_{total} in the whole frequency band is:

$$W_{total} = \left(W_{z1}^{\ 10} + W_{z2}^{\ 10} + \cdots + W_{zm}^{\ 10}\right)^{0.1} (m \le 4) \quad (16)$$

The stability performance decreases with the increase of the W_{total}, and the vehicle running stability is always evaluated to be in an excellent level when the value increases.

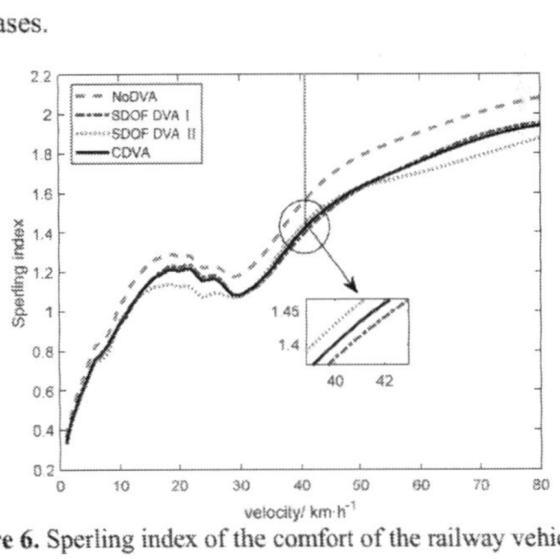

Figure 6. Sperling index of the comfort of the railway vehicle.

Fig. 6 is Sperling's riding index under no-load condition when the rail vehicle is equipped with CDVA. As can be seen from Fig. 6, with the increase of velocity, the vertical stability index of rail vehicles generally shows an upward trend and tends to be flat. When the vehicle is running under the conditions of unpowered DVA, installation of SDOF DVA and CDVA, the stability of the vehicle is kept in an excellent state. However, with the increase of the velocity of rail vehicles, the absolute value of the difference between the vertical stability indexes of the vehicles with or without the DVA shows an increasing trend, which indicates that the dynamic DVA has more room to improve the stability of the vehicles at medium and high velocity and the vibration reduction effect is more obvious. Compared with the SDOF DVA, the stability index of the CDVA is always in a relatively stable state, and generally better than that of the SDOF design, and there is no phenomenon that the stability index of the SDOF design fluctuates greatly up and down. Therefore, under the condition of full velocity, the stability index value of the vehicle is lower and more stable, which can effectively improve the running stability of the rail vehicle and the riding comfort of the vehicle. Especially in medium and high velocity operation, its vibration reduction effect is more significant.

7 Conclusions

(1) Through the vertical dynamic model of the carbody-CDVA, the vibration response of the carbody shows the coupling of bounce vibration and pitch vibration in low frequency band, and the acceleration PSD shows the enhancement and attenuation of vibration in different degrees. The SDOF DVA can suppress the above two kinds of vibrations of the carbody respectively, but once the design frequency is far from the peak frequency, the vibration reduction effect of the carbody is suppressed and there exists vibration increase within a partial scope.

(2) In order to ensure that the dynamic DVA can suppress bounce vibration and pitch vibration in low frequency band, the parameter design of the CDVA with double oscillator structure is proposed. By using the principle that the mass and design frequency of the two oscillators jointly act on the rigid vibration of the carbody, the CDVA can effectively control the multi-peak vibration in low frequency band of the vehicle.

(3) The vibration discreteness comprehensive index established by using the acceleration PSD at low, medium and high velocity to quantitatively evaluate the vibration damping effect of the dynamic DVA, the smaller the value, the smaller the dispersion degree of vehicle vibration, the smoother the vibration response of the vehicle, and the better the vibration damping effect. Under the same working condition, the damping effect of the CDVA is better than that of the SDOF DVA.

Acknowledgements

This research was partially supported by the National Natural Science Foundation of China (Grant No. 11472176) and the Natural Science Foundation of Shanghai (Grant No. 15ZR1419200).

References

1. Kazuto Seto 2013 *Dynamic vibration absorber and its application* (Beijing: China machine press)
2. Zhou J S 2012 *Railway vehicle vibration and control.* (Beijing: China railway publication house)
3. Zhou J S, Shen G, Zhang H and Luo R 2008 Application of modal parameters on ride quality improvement of railway vehicles. *Vehicle Sys Dyn.* **46(sup1)** 629–41
4. Foo E and Goodall R M 2000 Active suspension control of flexible- bodied railway vehicles using electro-hydraulic and electro- magnetic actuators. *Control Engineering Practice.* **8(5)**:507-518
5. Wu P B, Zeng J, Dai H Y 2004 Dynamic response analysis of railway passenger car with flexible carbody model based on semi-active suspensions. *Vehicle Sys Dyn.* **41** 774–83
6. Zeng J and Luo R 2007 Vibration analysis of railway passenger car systems by considering flexible carbody effect. *China Railway Society.* **29** 19–25
7. Shi H L, Luo R, Wu P B, Zeng J and Guo J Y 2014 Influence of equipment excitation on flexible carbody vibration of EMU. *Journal of Modern Transportation.* **22(4)** 195–205
8. Shi H L, Luo R, Wu P B, Zeng J and Guo J Y 2014 Application of DVA theory in vibration reduction of carbody with suspended equipment for high-velocity EMU. *Science China Technological Sciences.* **57(7)** 1425–38.
9. Wen Y P, Li Q, Shang H L and Xu X J 2016 A Study on vibration and transmission characteristics of long solid models under different frequencies. *J Vib Shock.* **35(21)** 53-62
10. Wen Y P, Sun Q, Zhou W H, Shang H L and Guo L S 2018 Modeling and simulation of vibration absorber for the carbody utilizing magnetorheological elastomers. *Chin J Mech Eng.* **54(22)** 114–27
11. Zhou W H, Wen Y P, Shang H L, Zong Z X and Guo L S 2018 Study on the vibration reduction of wide dynamic vibration absorber for the carbody utilizing magnetorheological elastomers. *J Vib Shock.* **37(16)** 124–34
12. Shen Y J, Wang X R, Yang S P and Xing H J 2016 Parameters optimization for a kind of dynamic vibration absorber with negative stiffness. *Mathematical Problems in Engineering.* **2016** 1-10
13. Shen Y J, Peng H B, Li X H and Yang S P 2017 Analytically optimal parameters of dynamic vibration absorber with negative stiffness. *Mechanical Systems & Signal Processing.* **85** 193–203
14. Tomioka T, Takigami T and Aida K 2009 Modal analysis of railway vehicle carbodies using a linear prediction model. *Syst Design Dyn.* **3(6)** 918–31
15. Tomioka T, Takigami T, Aida K 2012 *Conf. on Railway Technology.* p 18–20
16. Song M K, Noh H C and Choi C K 2003 A new three-dimensional finite element analysis model of high-velocity train-bridge interactions. *Engineering Structures.* **25(13)** 1611–26
17. Dumitriu M, Leu and Mihai C 2018 Correlation between ride comfort index and sperling's index for evaluation ride comfort in railway vehicles. *Applied Mechanics & Materials.* **880** 201–6
18. Munawir T I T, Samah A A A, Rosle M A A 2017 *Conf. on Materials Science and Engineering* vol **226(1)** p 12–39

Research on Maintenance Support System Based on Complex Network

Yang Ke [1,a], Fan Shi-dong [1] and Yao Yu-nan [1]
1 School of Energy and Power Engineering, Wuhan University of Technology, Wuhan, 430063, China.

Abstract. The design and collaboration analysis of maintenance support system network is a critical challenge. So the network model of ship maintenance support system was proposed, it involved all levels of distributed maintenance support participating nodes and formed a net chain structure. The network characteristic was analysed using complex network theory. The impact of topology characteristic and dynamic evolution on the cooperation effectiveness of maintenance support system was discussed. And an optimal evolution algorithm of support network was put up to so that the optimal support network would be designed.

1 Introduction

Maritime transport carries more than 80% of the international cargo traffic in the commodity exchange of economic globalization because of the advantages of capacity and cost. At the same time, the research on ship maintenance support based on operational reliability and economy has been widely concerned at home and abroad because of the competition in shipping market. More and more technology has been used to improve the reliability of state perception, which comes to a relatively accurate maintenance support prediction [1]. It is common for scholars to make a number of research on how to optimal the time window related to maintenance and support.

However, there has been a great gap from the maintenance demands to support response. Therefore the maintenance support supply chain network has become an effective means to improve maintenance benefits and operational collaboration.

Maintenance support system is driven by the maintenance requirements of ship equipment based on the supporting facilities. It controls at all levels of maintenance support resources dispersed in different locations through centralized or distributed information. Finally it forms an interrelated and closed collaboration system to achieve the supply, transport, allocation, collaboration and sharing of various types of resources[2]. It shows a high degree of networking in the geographical layout and operation relationship. Yin [3] abstracted the nodes of maintenance entities based on the flow and redistribution of elements in the maintenance implementation process. He established the topological meta-model and network relationship of maintenance support system. Through the change of statistical characteristic variables in the network structure, the influence of topology scale and information sharing and interaction ability on the efficiency of the support system is studied. Also the integrated mode of maintenance support is preliminarily discussed. Zhang [4] set up the perspective of complex network to study the command and control information synergy model. The synergy mechanism of uncertain supply and demand is studied from the view of case reasoning and Agent.

In general, the ship operation and maintenance supply chain is a complex network chain structure which revolves around the network nodes such as maintenance bases, suppliers, retailers, logistics enterprises and users. It mainly has the following characteristics:

1) Complexity. The operation and maintenance supply chain covers the entire process from raw material suppliers, parts suppliers, product manufacturers, retailers, service providers to terminal users. It is a large span structural pattern with multi-layer and multi-level participating entities, and enterprise nodes of different types, different geographies, different market positions, which all embody the complexity of maintenance supply chain network.

2） Dynamics. Firstly, node enterprises enter or leave the maintenance support network because of the random demand and the impact of the market environment. Secondly, due to the technological innovation and different business strategies of upstream and downstream node enterprises, the core

[a] Corresponding author: ake1231@163.com

enterprises change the supply chain cooperation relationship dynamically.

3） Self-organization. The maintenance support supply chain is composed of independent node enterprises such as core maintenance bases, resource suppliers, manufacturers and end users. They make continuous exchange of raw materials, energy, capital, labor, technical services and information with the outside world. There is competition and game between supply chain network node entities for individual interests. However, it finally comes to the co-operation on the basis of total network income by self-organization.

2 Construction and evolution analysis of ship maintenance support network model

The network research was first derived from the 20th century mathematician Euler's proof of the "Konigsberg seven Bridge Problem", which is typically characterized by the use of nodes and edges to abstract the operation of actual system. Nodes represent different individuals in the actual system, while edges represent a specific relationship between nodes (such as decision flow, information flow, logistics, capital flow, etc.)[5].

The onshore maintenance support entities involve the task ports, the supply stations, the repair shops, the shipyards, spare parts inventory at all levels and the suppliers in geographical dispersion. As a result, the network contains a series of sub-networks, such as organizational network, logistics supply network, inventory network, maintenance network, transportation network and so on.

As a result, the maintenance participating entities are divided into maintenance implementation layer and resource supply layer according to the attribute characteristics of the participating entities. Taking each maintenance support entity as the node, the maintenance support network model (shown in Fig.1) is abstracted from the geospatial layer on the edge of the interaction relationship (information flow, logistics, capital flow, etc.) between the maintenance supporting entities.

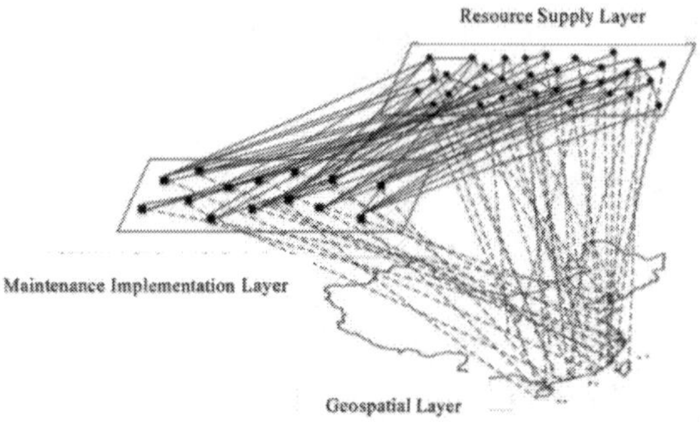

Fig.1 The maintenance support network model based on geospatial layer

The maintenance support network diagram is represented by G=(V,E), where $\mathbf{V} = \{v_1,v_2,...,v_n\}$ is a finite set of elements; $v_1,v_2,...,v_n$ represents the participating entity node in the maintenance support system; $\mathbf{E} = \{e_1,e_2,...,e_m\}$ is a finite edge set corresponding to the node pairs in set V, representing the interaction relationship based on different attributes[6].

In the network, with the change of time series, the dynamic evolution process of network is formed because of the addition and exit of network nodes. As an operation and maintenance supply chain network, the evolution behaviour of its nodes still follows the behaviour pattern of the general network structure: add（remove）the nodes, add（remove）the edges, re-connect. To some extent, these forms of evolution may exist at the same time.

Add a new vertex to the supply chain network, that is, the introduction of new manufacturers in the supply chain system and add a number of edges that depart from this vertex. Add edge refers to a number of new connections between the original vertices in the supply chain network, that is, the formation of a new supply or demand relationship in the supply chain system. Reconnect refers to breaking the original edge and then add another new edges to the same vertex, which means the manufacturer is excluded from the supply chain system and then it will re-join into the supply chain system. It is shown in Tab.1.

Table 1 The evolution model of supply chain network

Network representation	Evolutionary behavior and its significance
Add the nodes	New fundraising for node enterprises.
Remove the nodes	Old enterprise nodes were eliminated because they could not meet the needs of the network chain.
Add/remove the edges & Re-connect.	The change of node cooperative relationship

As a network of economic relations, supply chain network has obvious unequal time and preference for node arrival. In the process of supply chain network evolution, it is the source power of evolution for member enterprises with limited rationality to seek the optimization of their own enterprise profit. Therefore, the evolution of supply chain network shows a certain degree of concentration and randomness. The core vendors get most of the connections while other vendors get fewer connections because of the different status of nodes and the roles they assume in the supply chain. It is consistent with the actual supply chain system. For the existing manufacturers in system, the fluctuation of the manufacturer's state will also cause the departure of the original manufacturer, which is manifested as the break of the link and the detachment of the manufacturer in the supply chain network. When the manufacturer is out of the supply chain system, accordingly, the vendor node and its connection related to the node are deleted in the network.

It is assumed that the process of nodes joining and leaving supply chain system obeys the Poisson distribution. Therefore the rules for building maintenance network evolution are as follows [7]:

1) Given supply chain network system, it includes a number of m_0 enterprise nodes (t=0).

2) Growth: Per time step, there is a new node added to the network. It is connected with other existing **m** nodes. $(\mathbf{m} < m_0)$.

3) Connection priority: The probability that new node is connected to the existing node **i** is proportional to k_i. That is, the actual connection is affected by the node connection priority. $\prod_i = \frac{k_i}{\sum_j k_j}$.

3 Collaborative Efficiency Evaluation of maintenance supply chain network

Interaction between nodes is quote complex in the real marine maintenance support system. The traditional maintenance support focus on the local optimization of single node, however, the maintenance support system in complex network environment has a higher requirement on the information sharing and collaboration among every maintenance support unit. As a result, a moderate collaboration should be found in the premise that the reliability of maintenance could be ensured so that an optimizing strategy could be achieved for the operation and maintenance cost and system efficiency [8] [9].

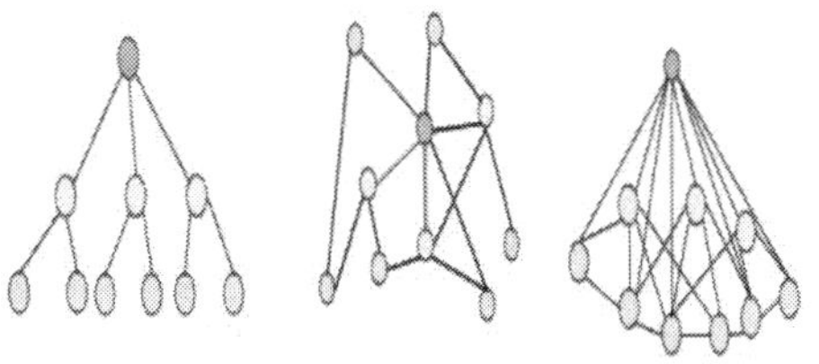

Fig.2 The maintenance support network synergy

In the traditional chained maintenance support system, the maintenance support entities lack collaborative relations. That is to say there is only superior and subordinate support relations among the entity nodes without the collaboration, information sharing and resource sharing. This kind of maintenance support system is low cost, small profit, and low efficiency. The opposite network has larger length of feature path and synchronized collaborative index, and low cost.

Moderate collaboration develops the sharing collaborative relations among limited information, resource and maintenance capability. The cost of the systems in this mode increases moderately, but the profit and efficiency increase significantly. The network characteristics in this mode has moderate length of feature path and synchronized collaborative index, and lower cost.

In excessive collaboration, all nodes in the maintenance support system tend to be in all connected mode based on the moderate collaboration. There is an interaction of information and resources between all maintenance support entities. The excessive collaboration maintenance network results in ultra-high costs without benefits improved accordingly. The network characteristics in this mode has lower length of feature path and maximum collaborative index, and higher cost.

For the maintenance support network, the synergy efficiency is mainly embodied in the following 3 aspects: the difficulty of the synergistic realization, the size of the synergistic cost, the advantages and disadvantages of the synergy effect [10]. Therefore, according to the definition of complex network, we should choose the appropriate performance index to establish the evaluation model of maintenance support network and calculate the overall topological evolution characteristics statistically. Then construct the best cooperative maintenance support network model. The evolutionary algorithm flow is shown in the figure 3:

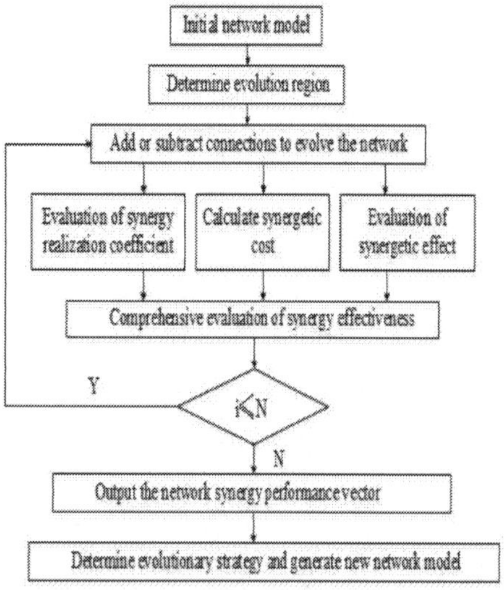

Fig.3 The maintenance support network evolutionary algorithm

1) On the basis of the initial network model, the network evolution area is determined.

2) In the evolutionary region, the evolution strategy is implemented in turn 1) to obtain the new network model and to calculate the collaborative efficiency under the new model. Then it generates the network synergy performance vector.

3) The region corresponding to the optimal value in network synergy performance vector is the optimal strategy of network collaborative evolution.

4) Iteration of the above process across the network, then global evolution strategy is determined and the collaborative performance curve in the process of optimal network evolution is calculated. Finally the support network with the greatest synergy efficiency is designed under the optimal collaborative evolution strategy.

4 Conclusion

Maintenance support is a solid foundation for the safe and efficient operation. The research of maintenance support network model based on supply chain perspective shows the core maintenance base and spare parts suppliers have obvious connection priority in the network structure and supply relationship. The complex characteristics of organization, self-adaptation and dynamic evolution affect the operation efficiency of the whole network to a certain extent. Also more research on supply network structure evolution and supply chain cooperative game based on collaboration will be developed.

References

1. Barabási A L, Albert R, Jeong H. Evolution of the social network of scientific collaborations [J]. Physical A,1999: 272-273.

2. Dekker R. Applications of maintenance optimization models: a review and analysis [J]. Reliability Engineering and System Safety, 1996, 51(3): 229 —240.

3. Xiaohu Yin. Research on the Dynamical Analysis Method for Equipment Maintenance Systems [D]. Hunan: Graduate School of National University of Defence Technology, 2008.

4. Tao Zhang. Modelling and Analysis of Maintenance Support Capability Assessment of Equipment in the Usage Phase. [D]. Hunan: Graduate School of National University of Defence Technology, 2004.

5. Sui Rubin，Zhong Guijuan. Research on networks of parts relation for product family based on complex networks [J].Commercial Research, 2009(1) :5053.

6. WANG Xiaofan, LI Xiang, CHEN Guanrong. Complex network theory and its application [M].Beijing: Tsinghua University Press, 2006.

7. Kwon O，Im G P，Lee K C. MACE-SCM: A multi-agent and case-based reasoning collaboration mechanism for supply chain management under supply and demand uncertainties [J].Expert Systems with Applications, 2007, 33(3) :690 -705.

8. GUO Lei ， XU Xiao-ming. Complex networks[M].Shanghai: Shanghai Science Education Press, 2006.

9. Xu Yuguo. Research on key Technologies for Equipment Autonomic Logistics [D]. Hunan: Graduate School of National University of Defense Technology, 2008.

10. Hanaki N, Peterhansl A, Dodds P S, et al. Cooperation in evolving social networks [J]. Management Science, 2007, 53(7):1036 —1050.

Comparing Road User Charging Acceptability in the City of Tunis and Damascus

Mohamad Shatanawi[1a], Souhir Boudhrioua[1] and Ferenc Mészáros[1]

[1]*Department of Transport Technology and Economics, Faculty of Transportation Engineering and Vehicle Engineering, Budapest University of Technology and Economics, Hungary*

Abstract. Worldwide, multiple studies have been trying to reduce traffic issues without physically changing the road network, this is when the congestion fees strategy has been considered as a favorable solution for the urban traffic issues. A fundamental condition that needs to be checked before the implementation of the road-pricing scheme is the acceptability of both the political and the public parties. The acceptability is so variable and depends on many features and differs from one individual to another, thus, a survey with a set of variant questions might help to understand the expectations and the worries of the citizens and aim to improve them for better effectiveness of the road-pricing project. This report aims, through analyzing the responses of a distributed survey, to evaluate the acceptability of the citizens of Tunis, Tunisia and Damascus, Syria in order to draw a comparison between the two cities. Moreover, it assesses the degree of acceptability and the variable expectations of the implementation of the congestion fees of the two societies.

1 Introduction

Over recent decades, the increasing number of passenger cars increased the additional social costs such as wear and tear of the road network and time lost due to traffic congestion. This reason behind this is, based on the economic approach, once a resource is underpriced or free, then the demand will definitely exceed the supply. Similarly, if the number of vehicles in a road exceeds its capacity, congestion is created.[1] To solve this issue, physical interferences on the road is not always possible [2] and they are just short-term solutions that might even lead to more congestion in the future[3]. Thus, the road user charging was introduced as a tool that showed its effectiveness in managing the traffic flows and consequently the travel demand, and in reducing traffic jams.

One of the earliest road toll application was in 1971, where only the road network has been considered[4], while other experiences relied on the measurement of the congestion's externalities and the optimal charges that can be allocated to tackle traffic jam[5]. Many cities started the implementation of this strategy such as London or Milan, yet many others still struggle with the socio-political feasibility of such projects like in Auckland.

In London for example, the recommended implementation of the congestion charging began in February 2003 as an answer to the issue[6] where the commuters spent 50% of their travel time in traffic queues daily in 2008 [7] and the congestion cost was 2 to 4 million pounds weekly. Despite the critics of several stakeholder groups such as drivers, business owners, and politicians, in one month the project succeeded to decrease the traffic jams, improve the bus and the taxi services with the revenues, the overall public acceptance increased [8].

In Milan, the cordon-based pricing scheme in the city center, was introduced between 2008 and 2011, Ecopass, to basically face the pollution with a small effect on the road users and/or improve the share of public transportation [7]. Thus, the price depended on the emission class of the vehicle. The scheme helped to meet the initial pollution mitigation objective than later turned to a congestion reduction approach in 2012[1]. It was highly accepted thanks to the different types of discounts [7] and an upgrade to expand the system was voted by the citizens with 79% of favor who also supported to include all types of vehicles[9].

Auckland had one of the highest car share, 555car/1000inhabitant since it is the most populated city of New Zeeland and has a low density and low public transport system [7]. Since 2005, researchers studied the acceptance of the congestion scheme but only 38% supported the project in 2005 and 25% in 2007[7]. The low public acceptance made the implementation of the scheme impossible to apply even after 13 years, which supports the idea that public acceptance is the key.

This paper studies the acceptability of road charging measures in two capitals, Tunis and Damascus, from two respective countries, Tunisia and Syria based on an online survey inspired the former survey[10], distributed

[a] Corresponding author: mshatanawi@edu.bme.hu

among the residents. Authors analyze at first the socio-economic variables, then traffic issues and their evolutions and the public and private acceptability at last. Finally, this helps in building a conceptual model targeting to improve the road charging acceptability. The parallel analysis aims to describe the differences and similarities between the two cities. The choice of the comparison between the two cities was made because of the similarities they share on different levels such as the socio-economic, the modal share of the transportation modes, the citizens' habits. Besides, both Tunisia and Damascus have a relatively rough political and economic situation.

The suggested congestion fees in the survey vary according to different time intervals for different zones in the city. Enforcement will be managed with cameras, which record all vehicle movements (enter or exit) in and out of the controlled zone. The proposed tariff strategy is the following as mentioned in Table 1:

Table 1. Toll Amount

Time	Congestion Fee	
	Tunis	Damascus
From 06:00 to 09:00	1.0 TND[1]	600 SYP[2]
From 09:00 to 15:00	0.5 TND	300 SYP
From 15:00 to 18:00	1.0 TND	600 SYP
From 18:00 to 21:00	0.5 TND	300 SYP
From 21:00 to 06:00	0.0	0.0

[1] 1 TND = 0.295 € (12 April 2019)
[2] 1 SYP = 0.001724 € (12 April 2019)

All the revenues stemming from the road pricing system will be used as mentioned below:

- 50% of the revenues will be spent on the development of the public transportation network,
- 50% of the revenues will be spent on the improvement of the road network.

2 Results and discussion

This section analyzes the different responses of the aforementioned survey distributed to 250 residents in Tunis and the same in Damascus.

2.1 Respondents analysis

In this part, the background of the participants, as a sample of the society, is analyzed. The gender repartition of the respondents is similar between in the two capitals, a valid percentage of about 55% male and 45% female in both cities. Most of the respondents in Tunis are between 21 and 30 years while in Damascus between 21 and 40 years.

Most people have a driving license with a higher percent in Tunis (88%) than in Damascus (78%), however, slightly more people in Damascus own a vehicle. Around 50% and 45% of the participants respectively in Tunis and Damascus use vehicles to commute to work/school, the second most used mode of transport in both cities is public transport (PuT), more commuters use it in Damascus. Few people walk or use bicycles for their trips, this reveals that the two cities are similar to the pattern of the transportation modes and that people are so dependent on their personal vehicles.

The analysis shows that having a vehicle influences the used mode of transport to commute to work/university where most of the respondents (80%) in both cities who have car use it for work or school trip. The behavior of the commuters is affected by the income and the occupational status of the respondents. Most of the participants are employees in both cities, followed by freelancers in Damascus and students in Tunis. More freelancers in Damascus might justify the higher number of people owning a vehicle in Damascus

The analysis of the different features of the participants, have similarities in the citizens' behavior and their social background, which makes it easier to compare the acceptability between the two cities towards the implementation of the road charging scheme.

2.2 Traffic problems analysis

This part of the survey presents a set of questions aiming to determine the severity of the different traffic issues from the point of view of the participants.

The problems derived from the traffic can be classified into environmental problems (noise, air pollution, etc.), and problems directly related to the circulation of the traffic (Other problems). These two categories are shown in Figure 1. The hypothesis here is that if the respondents perceive these as a major or serious problem then they will likely be more ready to accept the pricing as a measure to solve these problems.

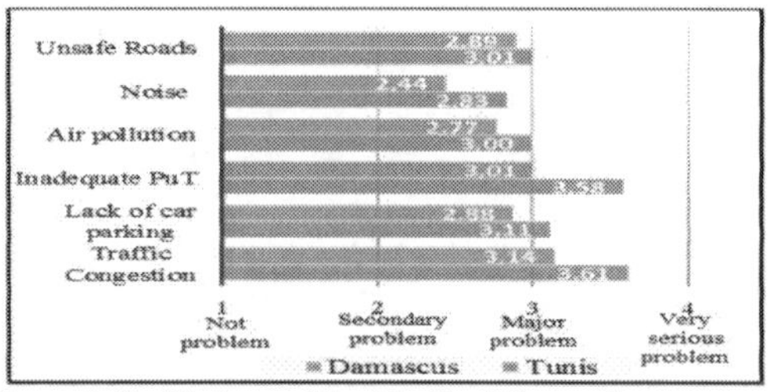

Figure 1. Traffic Issues Classification/ Mean values

The analysis shows that the awareness of the traffic problems is higher in Tunis, in both categories. However, in Damascus as well the problems are considered as major ones. Tunis's main problems are (1) the traffic congestion and (2) the inadequate and inefficient public transport (PuT) which are the main problems in Damascus. However, the problems mentioned above are considered serious in Tunis while they are major ones in Damascus. About the awareness of environmental problems, they are all considered major problems, but they are considered more serious in Tunis than in Damascus. There are many parties responsible when it comes to the issue of traffic. Figure 2 shows the responses' mean values regarding this matter.

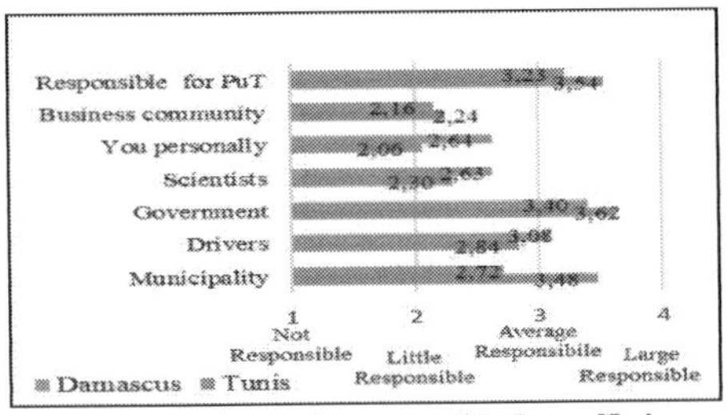

Figure 2. The parties responsible for traffic issues

Figure 2 explains that, in both cities, the respondents consider the authorities are responsible for the traffic problems, however, they feel more responsible in Tunis than in Damascus. The acceptability level is affected by whether respondents consider themselves as victims or contributors to traffic problems.

The last question was to understand the importance of different travel habits and whether they want to use their vehicles anytime and anywhere or they prefer other transport modes. Surprisingly the mean values of answers show high interest in all options. This contradicts can be interrupted as the respondents feel the necessity to use their vehicles due to the absence of the adequate alternative but hope for mobility development in the future.

2.3 Study of the acceptance of the proposed scheme

After the introduction of the scheme, the set of questions aim to capture preliminary opinions about the road charging concepts, such as the effectiveness, the effects on their mobility and travel behavior, the acceptance and the usage of raised fund.

The first studied point is the knowledge about the scheme by asking the participants to which extent they have heard about the pricing concept, Figure 3. Having information about the benefits of congestion pricing may increase the acceptability for, otherwise, it might be seen as a limitation of one's liberty.

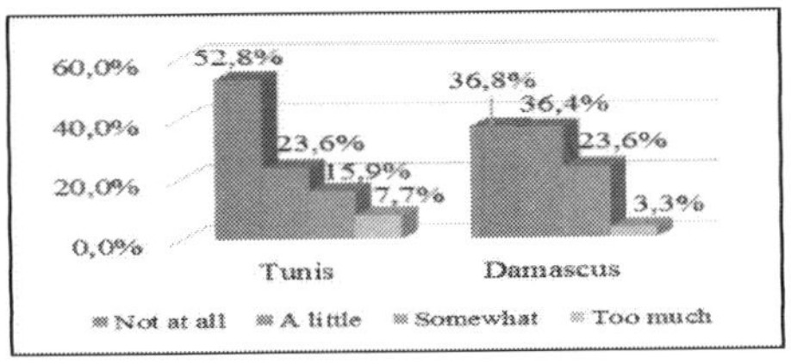

Figure 3. Knowledge about the proposed scheme

More than 70% of the respondents in both cities have not heard about road charging scheme or have just a little idea, which indicates the lack of information regarding the scheme in both cities.

The following question discuss the expected effectiveness of the road-charging scheme in reducing the traffic amount in both cities. If the respondents believe this measure is effective then their acceptance will be higher. A high expectation for the scheme effectiveness in Tunis as almost 65.7% of the respondents believe this measure will reduce the traffic amount in the city, while in Damascus half people think it could affect and other think it will not affect the traffic.

The acceptance for implementing the scheme is asked in the survey in two questions (1) about the public acceptability and (2) about the acceptability of his/ her family and close friends, individual acceptability

Answers are almost similar in both cities where slightly more than half respondents think that it would not be accepted by the public, where others believe the public will accept the measure. However, the percentage of those who think it is a completely acceptable strategy for the public is very small and those thinking it is totally unacceptable is forming a significant percent, which leads to considering higher refusal to the proposed scheme. Surprisingly, a higher level of acceptability can be seen when it comes to the expectation of the family and close friend's acceptability 62.6% in Tunis, 63.4% in Damascus.

The scheme may affect the several traveling habits of the daily commuters where the next question discusses these possible changes in case the scheme is implemented, and the provided questions can be grouped into two factors. The first is the positive changes by reducing the personal car trips and the second factor is the negative where the commuters will keep using the personal car with a slight shift in the times or the routes. Answers were based on a scale from 1 to 4 (where 1 is "No, for sure" and 4 is "Yes, for sure") and the mean results of all variable as seen in Table 2.

Table 2. Scheme Effects on the commuter's habits

	Tunis	Damascus
I will drive the car less	2.55	2.48
I will use more the public transport	2.53	2.63
I will use the bike more or walk more	2.64	2.45
I will pay the new fees and continue driving as before	2.33	2.38
I will not drive in tolled areas	2.53	2.17
I will share my car with my colleagues and vice versa	2.40	2.42
I will not drive during the charging times	2.45	2.18
I will support the demonstrations to stop the project	1.54	1.66

Again, the answers are similar in both cities and the proposed scheme will not affect the travel habits significantly. The participants are moderate in their responses and there is no wish to protest. They would most likely either use public transport, bike or walk more.

The dependency on using the personal vehicle is an important factor in the acceptability for the road fees; the commuters who can reduce their trips with a personal car are more likely to accept the measure. The study shows a higher dependency on using a personal car in Damascus with 52.7% of people who disagree to significantly decrease the use of their vehicles.

The respondents' anticipation for the elements affecting their trips are analyzed here, these elements are divided into two factors (1) favorable factor- respondents anticipate good results of implementing the scheme- and (2) unfavorable factors where the respondents anticipate bad results. The idea here if factor (1) is more anticipated by the respondents then the scheme will have more acceptability. The mean value of the answers which were designed on a scale 1 to 4 (1 is "No, for sure" and 4 is "Yes, for sure") are shown in Table 3.

The Tunisians respondents expect the proposed scheme to make the city a better place and reduce the travel time and environmental problems. Yet, Syrians in Damascus, the estimations are lower and additional travel costs are expected.

Table 3. The estimated consequences of the proposed scheme

	Tunis	Damascus
Shorter travel time	3.10	2.68
Additional travel costs	2.48	2.71
Less environmental problems	3.06	2.64
Extra effort in planning trip	2.74	2.52
A nicer area (fewer vehicles)	3.20	2.62
Restricting my travel unfairly	2.26	2.42

The generated revenues from applying road charging scheme shall be used in benefits of the road users and the citizens, when asked how they shall be used and how might the government distribute them, the participants in Tunis highly agree to use them to improve the PuT, 93.3% and the movement of pedestrians and bikers, 91.2%. Yet, they do not expect that the government would do so; but it would probably support the budget of the state or improve the traffic network such as roads. Contrarily to Tunis, in Damascus, there were no extreme choices, only for the case of the agreement on improving the PuT with 84.4% and unlike Tunis, the respondents agree to support of the budget of the state with 63.6%. Similarly, to Tunis, the respondents disagree to decrease the customs fees on the vehicle and estimate that the government is likely to improve the infrastructure and support its budget.

2.4 Multivariate Analysis

This section discusses the further multivariate analysis that was used to examine what affects the acceptability of the scheme. In this manner, we used the dimension reduction technique or factor analysis "FA".

2.4.1 Sensing the problems

In this question (Figure 1) the aim is to evaluate the degree of sensing the traffic problems, the Cronbach Alpha test, for the six items, with an alpha coefficient of 0.72, suggesting that the items have high reliability. Moreover, the Kaiser-Meyer-Olkin (KMO) test of 0.741 shows the sampling adequacy for the factor analysis of this question. There were two Eigenvalues greater than 1, hence two factors are retrieved explaining more than 60% of the total Variance of the data. The rotated component matrix shows which variables have the highest loadings within each factor. Using the Varimax rotation method guarantee that the variable can have a high loading in exactly one factor.

Accordingly, the process of simplifying and naming each factor become clear.

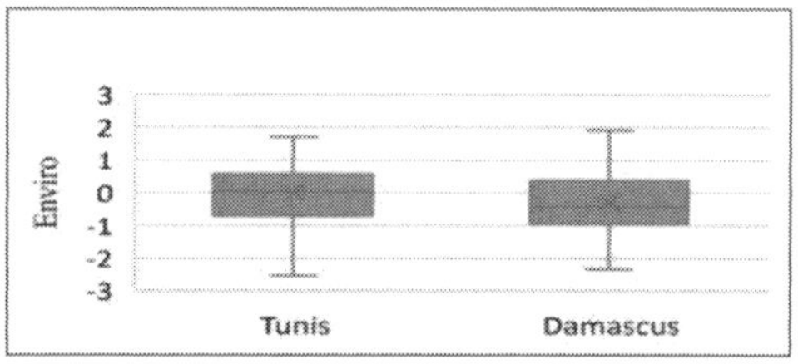

Figure 4. Problem sensing/ Environmental factor

The results retrieved two factors as follow (1) Environmental factor, which includes air pollution, traffic noise, and unsafe road, and (2) Service factor which includes traffic congestion, lack of adequate parking places, and inadequate public transport system. Figure 4 and Figure 5 displays the distribution of the factor scores of these two factors in both capitals, Tunis and Damascus.

The Boxplot shows a higher interest in environmental issues in Tunis compare to Damascus. The spread of the answers is approximately similar in both cities. However, there is a negative skewness in Tunis and positive skewness in the answers of Damascus which explain the wider range of responses in Tunis where there is a low percent consider the traffic-related problems as secondary problems.

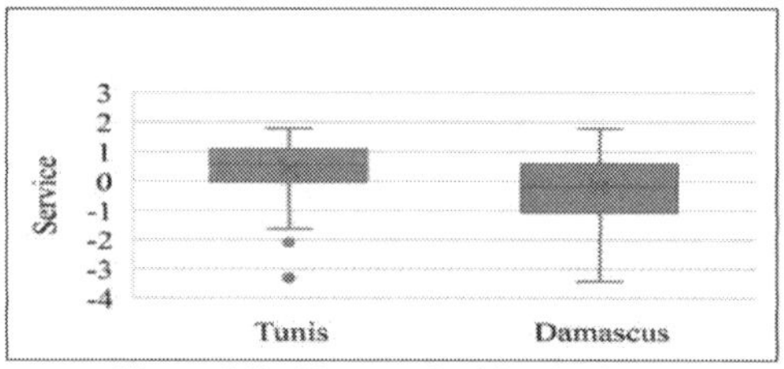

Figure 5. Problem sensing/ Service factor

Again, the median regarding the service factor is higher in Tunis compared to Damascus, which shows higher sensing for the problems in Tunis. As well as the answers look more consistent in Tunis with the existence of outliers where these outliers show the respondents who said the mentioned problems are not a problem at all or just secondary problems in Tunis. The skewness in both capitals is negative as fewer number of the respondents considered the traffic-related problems as secondary ones.

2.4.2 Parties responsible for traffic issues

The question which discusses the responsible for the traffic problems (Figure 2) consists of 7 elements, combining these elements gives true score variance with Cronbach's alpha of 0.688. Factor analysis is possible as KMO test value is 0.744 and the Total Variance Explained table present two factors (Eigenvalue >1) where those two variables can be called as "Government" and "People" and they explain more than 55% of all variables. Most Tunisians blame the government for the traffic problems, while opinions in Damascus are more moderate and with wider range, the reverse appeared when the second factor "People" was considered as it is seen in Figure 6, the Syrian believe they are more responsible for the traffic problems than Tunisians do, with smaller range and more consistency in the answers.

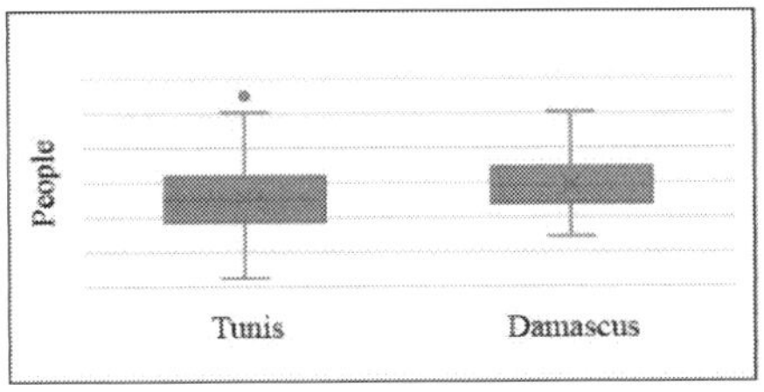

Figure 6. People responsibility for traffic problems according to respondents

2.4.3 Acceptability

There are two questions measure the direct acceptability - the respondent's point of view for the acceptability of the society in general and the acceptability of their close friends and family member - these two questions were gathered, and boxplot was created to compare the two cities acceptability. Surprisingly, Figure 7 show very high similarity in both cities and with approximately similar median, minimum, maximum, and range.

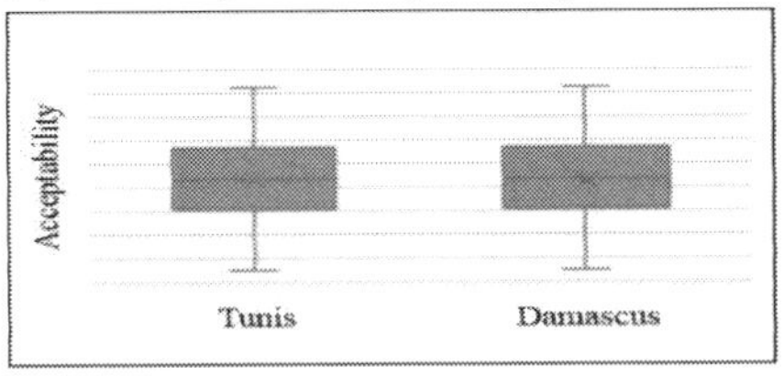

Figure 7. Acceptability of the scheme

2.4.4 Anticipation for the trip elements after implementing the scheme

The six elements in table 3 were gathered and tested, the Cronbach alpha is 0.585 for all elements where the KMO test gives a result of 0.670 and the total variance explained table suggests two factors which explain more than 63% of all variables. The rotated complement matrix gives two factors (1) favorable factor which includes item number 1, 3, and 5, Table 4, (2) unfavorable factors which include item number 2, 4, and 6 in the same table.

No.	Items	Factor loading	
		1	2
1	Shorter travel time	0.811	-0.065
2	Additional travel costs	0.001	0.740
3	Less environmental problems	0.855	0.047
4	Extra effort in planning trip	-0.178	0.794
5	A nicer area (less vehicles)	0.823	0.071
6	Restricting my travel unfairly	0.240	0.672

Table 4. Rotated component matrix

The positive expectation in Tunis is much higher than in Damascus in case the scheme was implemented, with more consistency in the answers and smaller range. The outliers show the few numbers who expected no good effect at all after implementing the scheme. On the other hand, the negative expectation is higher in Damascus and more consistency with a smaller range.

3 Future Work

This research is being expanded to include on one hand more cities in order to evaluate the acceptability conditions in Budapest, Ulaanbaatar, and Amman. On another hand, an application of the factor analysis to all questions of the survey shall be used as well as the regression analysis and other statistical methods to find the influencing factors on the acceptability of road charging scheme. Thus, the gathering would help create a conceptual model to improve the acceptability of the road charging scheme and to be used as reference or cornerstone.

4 Conclusion

Road charging scheme can be used as a measure to solve or reduce traffic-related issues by modifying the traveling behavior of the commuters. However, it is very difficult to convince the public to pay new fees for using roads where they did not use to pay for this before that is referred to as "low acceptability". This can be enhanced by using the right measures such as finding alternatives or better allocation of the revenue, etc. This paper focuses on the acceptability of road charging scheme in two capital cities (Tunis and Damascus). Although the two societies have similarities, the results show a different attitude towards traveling behavior and perspectives.

In both cities, people consider the traffic related issues as major problems with focusing on the service factor more than the environmental factor. Moreover, Tunisians are more frustrated with bad future expectation and blame the government for these problems, while Syrians see better future traffic situation and surprisingly, see themselves as a society more responsible. Moreover, Syrians have a better knowledge of the scheme. However, Tunisians have a more optimistic opinion for better mobility situation in case the scheme was applied to the city. These differences between the two societies dealing with different factors which affect the acceptability produce similar acceptability level in both capitals.

References

1. A. E. Ohazulike, E. C. Van Berkum, G. Still, and M. Bliemer, "Multi-objective road pricing: A cooperative and competitive bilevel optimization approach," p. 6, (Nov. 2010).
2. Gumz, F. and Török, Ádám. "Investigation of Cordon Pricing in Budakeszi", Periodica Polytechnica Transportation Engineering, vol. 43, no. 2, pp. 92-97, (2015)

https://doi.org/10.1051/matecconf/201929602002

3. D. Pojani and D. Stead, "Sustainable Urban Transport in the Developing World: Beyond Megacities," *Sustainability*, vol. 7, no. 6, pp. 7784–7805, (Jun. 2015).
4. Dafermos, S.C., Sparrow, F.T., 1971. Optimal resource allocation and toll patterns in user optimized transport network". J. Transport Econ. Pol. 5, 198–200
5. Pigou, A.C., The economics of welfare. Transaction Publishers (1924)
6. T. Litman, "Congestion Pricing, Implications for Other Cities," *Vic. Transp. Policy Institue*, p. 14, (Nov. 2011).
7. V. Jarl, "Congestion pricing in urban areas: theory and case studies," Lund Institute of Technology, Department of Technology and Society, (2009).
8. T. Litman, "London congestion pricing-implications for other cities," *Ifo Inst. Econ. Res. Munich*, pp. 17–21, (2005).
9. G. Mattioli, M. Boffi, and M. Colleoni, "Milan's pollution charge: sustainable transport and the politics of evidence," *White Rose Research*, p. 17
10. J. Schade and B. Schlag, *Acceptability of urban transport pricing*. Helsinki: Valtion Taloudellinen Tutkimuskeskus, (2000).

Research on Two-stage Order Picking Sequencing for Intensive Shelf

Xue Tian [1,a], Li Zhou [2] and Jianglong Yang[3]

[1]*Associate Professor of Beijing Wuzi University, China; Visiting Assistant Professor in CSUSB, California State University, San Bernardino, USA*
[2]*Professor of Beijing Wuzi University*
[3]*Student of Capital University of Economics and Business*

Abstract. The increasingly usage of Intensive shelves, greatly increase the utilization of storage space, but also is more demanding on order picking time . Based on the storage layout of intensive mobile shelves, this paper combines the time cost of shelf movement with the moving distance under the guidance of seeking global optimum. Transform the single order picking process into TSP problem, while considering the picking process. The waiting cost of the order, and the minimum of the picking cost and the waiting cost of the whole batch of orders requiring shelf movement, to establish a two-stage mathematical model of the order picking order. Then, the algorithm for solving the model is designed, and the simulation is carried out by numerical examples to illustrate the law and characteristics of the problem more vividly, in order to provide reference and reference for the order picking activities of intensive mobile shelves.

1 Introduction

In the current era, with the continuous development of e-commerce and urban distribution, the efficiency of logistics activities has also been more and more important. As a core function of logistics activities, warehouse activities have attracted much attention. As far as order picking in warehouse activities is concerned, it has long been recognized as one of the most time-consuming activities in warehouse activities, and is also an important bottleneck for improving the efficiency of warehousing activities. At the same time, due to the increasingly scarce characteristics of land resources in urban areas, the storage cost of goods is rising rapidly, which requires the storage area to increase the utilization of warehouse volume as much as possible. As a result, a more intensive storage shelf layout has emerged in which intensive mobile shelves have emerged and are increasingly being applied to the practice of warehouse activities. Although the intensive mobile shelf can improve the utilization of the storage space of the warehouse, it needs to remove the goods by continuously moving the shelves during the order picking process, which in turn further aggravates the time consuming of the order picking activities, even the shelf moving place. It takes much longer than the time to take the goods out of the shelf, resulting in a reduction in the efficiency of the storage operation to some extent.

Despite this, in some cases (for example, the rent of the warehouse location is much higher than the time cost), the use of intensive mobile shelves for warehouse layout to improve inventory space utilization is still the choice of some enterprises. Then, how to further improving order picking and improve picking efficiency has become an important issue for enterprises to think about.

2 Introduction to the status of intensive mobile shelves

2.1 Intensive mobile shelves use status

2.1.1 Introduction to intensive mobile shelves

Intensive mobile shelves, also known as dense shelves or dense cabinets, are densely stored cabinets or shelves. Because they are removable and densely arranged, they can save a lot of storage space, as shown in Figure 1.

Figure 1 intensive shelf

The intensive mobile shelves are arranged according to the characteristics of the warehouse, and the space used. The multi-section and multi-connection are located on a fixed track to form a whole body. The columns are

[a] Corresponding author: xue.tian@csusb.edu

accessed by hand or electric, and the operation is easy and flexible, and the operation is stable[1]. The contact surfaces between the dense mobile shelf frame is cushioned and magnetically sealed. There is a dustproof plate at the top and a mouse-proof device at the bottom. It is good dust, mouse, moisture and fire resistance.

2.1.2 Intensive mobile shelf structure

Generally speaking, the shelves are generally composed of cabinets, rails, transmission systems, etc. (see Figure 1), as follows:

① cabinet body includes: column, partition, hanging plate, hanging piece, side plate, top plate, dividing rod, bottom frame, door panel;

② transmission system includes: bottom rail bearing, cast iron angle wheel, transmission chain, transmission wheel, brake steering wheel or rocker;

③ Tracks include: rail seat, solid core steel, and anti-dumping device at the track end;

④ Other accessories: seal strip between columns, dust plate at the top, brake switch, total lock, label holder, etc.

2.1.3 Intensive mobile shelf types

① Classification by storage items: intensive archives, dense file shelves, dense data shelves, dense map bookshelves, dense tape racks, intensive medical records, dense blueprints, dense base frames, dense specimen racks, and dense sample racks;

② Classification by storage location: archive room intensive cabinet, data room intensive cabinet, library room intensive cabinet, storage room intensive cabinet, archive intensive cabinet, data warehouse intensive cabinet, warehouse intensive cabinet;

③ Classified by unit of use: archives intensive cabinets, hospital intensive cabinets, densely populated cabinets, enterprise intensive cabinets, school intensive cabinets, design institute intensive cabinets, insurance company intensive cabinets, bank intensive cabinets, supermarket intensive cabinets;

④ Classification by the production process: no door dense cabinet, semi-closed dense cabinet, fully enclosed dense cabinet, two column type dense cabinet, three column type dense cabinet, box type dense cabinet.

2.2 Research status of intensive mobile shelves

2.2.1 Research on optimization strategy of warehouse picking system

The existing warehouse picking mode can improve the flexibility of warehouse to a certain extent, but there are inevitably some problems with various types of warehousing picking modes, which has led to some scholars' relevant research.

BoysenN et al. (2017) explored the problem of order picking queues for mobile shelf warehouses, and reduced the time required to move heavy shelves by optimizing the order queueing, and enhanced picking efficiency; HeZ et al. (2018) in the context of intelligent warehouse automation services, provided a differentiated probability queuing strategy for serving customer orders using the AGV system, which enables to solve the joint optimization problem and reduced the processing time by 19.64%; RoyD et al. (2019) established the mobile fulfillment system based on the multi-class closed queuing network model.

2.2.2The development of the flexible storage system

The development of the warehousing system highlighted features such like compact, intelligent, economic, efficient, stable, and easy to expand[2]. The warehouse layout is developed from wide-channel g to narrow-channel, non-channel-to-person picking mode. the selection of picking equipment is changed from forklift, hydraulic handling, paper guiding to sound and light guidance, electronic label, intelligent conveyor belt, AGV intelligent equipment etc.

3 Intensive mobile shelf picking problem

3.1 Warehousing layout assumptions

Suppose there is a warehouse with a set of mobile dense shelves. As shown in Figure 2, the aisles between each rows of shelves can be opened, indicated as A_i $(i = 1,\ 2,\ \ldots\ldots n)$ 。

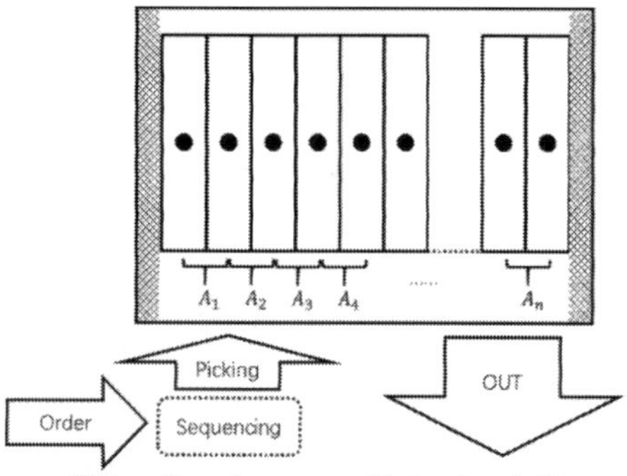

Figure 2 work process of intensive shelf

The time taken to open each aisle is represented by D_{ij} (i, j = 0, 2, ... n), which is the time it takes for opening the aisle from the i position to the j position. Assuming that the time cost per channel of moving a row of shelves is v, the time cost of all the moving times can be expressed as a table related to v, see Table 1.

Table 1 cost of all the moving times

D_{ij}		j				
		1	2	3	…	n
i	1	0	v	2v	…	(n-1)v
	2	v	0	v	…	(n-2)v
	3	v	v	0	…	(n-3)v
	…	…	…	…	…	…
	n	(n-1)v	(n-2)v	(n-3)v	…	0

3.2 Order picking hypothesis

Assume we need to pick a batch of orders within a certain time requirement. the goods picking time is affected by the moving shelf status and the order contents[3]. the entire batch of orders is already sorted and the picker is arranged to pick the goods with a certain sequencing. there is only one picker and one mobile channel in our model.

Suppose that in a certain period, the number of orders received is m, and B is used to represent the order, then there are $B=\{B_1, B_2, \cdots, B_m\}$. Concerned the goods in each order and location of each shelf, the aisle needs to be opened can be determined. That is, the mobile aisle corresponding to the order B_d needs to be opened, which can be derived from the following collection:

$$(E_{d1}, E_{d2}, \cdots, E_{dn})$$

Where E_{dj} = 0 or 1; d = 1, 2, ... m; j = 1, 2, ... n. When $E_{dj}=0$ the related aisle A_j does not need to be opened; when $E_{dj}=1$, the related aisle A_j needs to be opened. See Table 2 for the mobile aisle that need to be opened for m orders in the certain period.

Table 2 mobile aisle need to be opened

	A_1	A_2	……	A_n
B_1	E_{11}	E_{12}	……	E_{1n}
B_2	E_{21}	E_{22}	……	E_{2n}
……	……	……	……	……
B_m	E_{m1}	E_{m2}	……	E_{mn}

Thus, the relationship between the aisle that need to be opened by multiple orders can be further analyzed. thereby coordinating the ordering arrangement of all the orders in the entire on period.

4 Two-stage model of intensive mobile shelf order picking

Since the time cost of moving shelf is much higher than the picking cost, in the model picking time is not considered. Also, the orders follow a FCFS rules, so the later orders will generate waiting costs before is actually picked[4]. The longer the waiting time cost, the lower customer service level. Therefore, when establishing the model, it is necessary to consider both the time cost of moving the shelf and the time cost of order waiting[5]. A comprehensive balance is required to minimize the sum of the total time cost.

4.1 Stage 1: Finding the minimum time cost for a single order

First consider the time cost of a single order B_d . From the previous assumptions it is clarified the time cost of the aisle movement is linearly and positively related to the moving distance. Therefore, each aisle needs to be opened can be regarded as a point, and the whole process of picking for order B can be token as a TSP problem. We use Y_n to stand for the location of each shelf. As shown in Figure 3, $A_1, A_2 \cdots A_n$ are different location of the shelves, and the specific distances between any two shelves are shown in Table 1. The initial position of the moving channel is A_{j^0}. In addition, let A_0 be a virtual point, the distance from A_0 to the initial position A_{j^0} is 0. The distance from A_0 to other points is equal to the distance from the initial position A_{j^0} to other point, and it is 0. In addition, since each order needs to open different channels, if an order needs to open all channels, its network diagram is the whole picture of Figure 3. If an order only needs to open some channels, its network diagram is just part of Figure 3. Therefore, when the channel needs to be opened according to different orders is determined (see Table 1), the network diagram can also be determined.

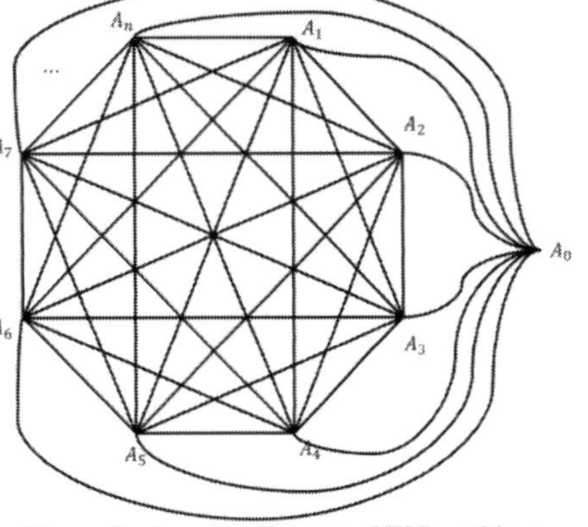

Figure 3 schematic diagram of TSP problem

Let F_{ij} indicate whether to move from channel A_i to channel A_j according to order B_d. D_{ij} stands for the cost per unit distance needs to be moved. Then, under the previous assumptions, we could establish a model like the following which shows that the target of the single stage optimization is lowest time cost.

$$F_{ij}=\begin{cases}1 & \text{from i to j}\\ 0 & \text{no need from i to j}\end{cases} \quad \text{where, } i, j=1, 2\cdots n$$

$$\min Z_d=\sum_{i=1}^{n}\sum_{j=1}^{n} D_{ij}F_{ij}$$

$$s.t.\begin{cases}\sum_{i=1}^{n} F_{ij}=1 & j=1, 2, \cdots n\\ \sum_{j=1}^{n} F_{ij}=1 & i=1, 2, \cdots n\\ F_{ij}+F_{ji}\le 1 & i\ne j\\ F_{ij}+F_{jk}+F_{ki}\le 2 & i\ne k\ne j\\ \vdots & \\ F_{ij}+F_{jk}+F_{kl}+\cdots+F_{pi}\le n-1 & \\ \quad i\ne k\ne\cdots\ne p\ne j & \\ F_{ij}=0 \text{ or } 1 & i, j=1, 2\cdots n\end{cases}$$

4.2 Stage 2: find the minimum time cost of a batch of orders

After the start of the picking process, what the waiting cost of the order, which has not been picked up, needs to increase is the time taken for the previous order picking. The d_s indicates that the order numbered d is picked at the S_{th} time, Z_{d_s} indicates the picking time of the order B_d.

It can be seen that after the picking of an order is completed in the S_{th} time, the sum of the waiting costs of the remaining (m-s) orders is increased by $(m-s)Z_{d_s}$. To complete m order picking during the entire picking process, the sum of all order waiting costs is $\sum_{s=1}^{m} (m-s)\, Z_{d_s}$.

The overall goal of this problem is to find that the sum of the picking cost and the waiting cost of all the picking orders in the inspection cycle is the smallest.

$$\min W = \sum_{s=1}^{m} (m-s+1)\, Z_{d_s}$$

$$\begin{cases} \forall d_s \in Q - \bigcup_{i=1}^{s-1} k_{i-1} \\ Q = \{1,\ 2,\ \ldots\ldots m\} \\ k_0 = \emptyset \\ s = 1,\ 2,\ \ldots\ldots m \end{cases}$$

Therefore, through the construction of the two-stage model, it is finally possible to obtain the global optimal solution, with the smallest sum of the order picking cost and the waiting cost, under the intensive mobile shelf of a single moving channel.

5 Two-stage model algorithm design for order picking

From the previous analysis, we can see the minimum picking time cost problem under our assumption is a NP hard problem[6]. That means it is generally difficult to find an accurate optimal solution. However, the genetic algorithm combined with Matlab software could helpful to solve this problem[7]. The specific algorithm is designed as follow:

5.1 Coding

(1) we first randomly sequence a mobile channel that needs to be opened. Then take this sequencing as a gene. For example, sequencing the channels that need to be opened according to order B_d, could be transformed as, sequencing all the non 0 variables in the set $\{E_{d1},\ E_{d2},\ \cdots,\ E_{dn}\}$, which is a gene;

(2) On the basis of the previous step, we could get *m* genes corresponding to *m* orders .

5.2 Generating an initial population

As shown in Table 3, the initial population is generated by decimal coding, and there is a total of p individuals, each of which has *m* genes, and each gene has a number of elements between 0 and n.

Table 3 the initial population

serial number	Picking order
1	{ (1, 3...) ,(2, 5, ...), ...}
2	{ (2, 8...) ,(6, 9, ...), ...}
......	......
p	{ (7, 2...) ,(1, 5, ...), ...}

5.3 Determine the fitness function

The fitness function at this stage is:

$$W = \sum_{s=1}^{m} (m-s+1)\, Z_{d_s}$$

Calculate the size of the individual's fitness, that is, the time spent on each order picking in the order in which the different channels are opened.

5.4 Individual evaluation: fitness check

The fitness of p individuals was calculated in turn, as shown in Table 4.

Table 4 fitness of individuals

Individual serial number	1	2	3	......	p
Adaptability	W_1	W_2	W_3	......	W_p

5.5 Copy, transfer, variation

(1) Arranging the various project according to the serial number, we can find the value of the fitness of each individual[8]. According to the order of the individual serial numbers, the cumulative value K_c of the obtained fitness can be calculated as $K_c = W_1 + W_2 + \cdots + W_p$. Meanwhile, as shown in Table 5, random number R_c between 0 and $K_p$could be generated and the total number of R_c is p (p is a positive integer large enough) [9].

Table 5 variable of copy, cross and variation

serial number (1≤c≤p)	1	2	3	4	5	...	p
Adaptability: W	W_1	W_2	W_3	W_4	W_5	...	W_p
Accumulated value of fitness: K_c	K_1	K_2	K_3	K_4	K_5	...	K_p
Random number between 0 and K_p: R_c	R_1	R_2	R_3	R_4	R_5	...	R_p

The individual copy rules are as follows:

① Compare the values of R_1 with the values from K_1 to K_p , one by one. The first K_c that is greater than R_1 is selected, and the corresponding individual serial number is copied once;

② Compare the values of R_2 with the values from K_1 to K_p , one by one. The first K_c encountered greater than R_2 is selected, and the corresponding individual serial number is copied once;

③ and so on, until the number of individuals copied reaches p, then the copy is stopped.

(2) Do non-put back random sampling until we got the schemes numbers more the 0.3 times *m*, then do the following operations:

① randomly select two elements inside each gene in each sorting scheme, and exchange the pair of elements positions ;

②Randomly select a picking scheme, randomly select two gene in this scheme, and exchange the positions of the two genes.

③ Until all the selected schemes have been done.

5.6 Termination conditions

The termination condition is 50 times. We need to find the position where the fitness function value is optimal and stable and take this solution as the optimal solution of the problem[10].

6. simulation samples

There is a set of 30 rows movable shelves, which has one moving channel that can be opened . Its initial state is opened at Y_5. There is a batch of 50 orders, and the specific channel conditions that each order needs to open are represented by a randomly generated 50×30 0-1 matrix. The sum of the position distance of the moving channel and the waiting cost of each order will be selected as the observed variables[11]. Based on Matlab software, the genetic algorithm program is written for this problem, and the target is to find optimal value with a lowest time cost[12]. The specific procedure is like the following:

Intensive mobile shelf two-stage order picking model genetic algorithm program

```
m = 50;n = 30;g = 20;ga = 500;T0 = 5;B = [];W = [];
D = cell(1,20);E = cell(1,20);JL = cell(1,g*ga);
A = randi([0,1],m,n);
for i = 1:m
| B = [B;([1:n])];
end
C = A.*B;
for j = 1:g
| for i = 1:m
| | temp = randperm(n); C(i,:) = C(i,temp);
| end
| D{1,j} = C(randperm(m),:);
end
ZD=zeros(1,ga);ZX=zeros(1,ga);
JZ=zeros(1,ga);ZS=zeros(1,ga);
for gacl = 1:ga
| for s = 1:g
| | Z = 0;
| | for j = 1:m
| | | for i = 1:n
| | | | if D{1,s}(j,i)>0
| | | | | Z = Z+abs(D{1,1}(j,i)-T0);
| | | | | T0 = D{1,1}(j,i);
| | | | end
| | | end
| | end
| | W = [W,(m-j+1)*Z]; JL{1,g*(gacl-1)+s} = D{1,s};
| end
| ZD(gacl)=max(W); ZX(gacl)=min(W);
| JZ(gacl)=mean(W); ZS(gacl)=mode(W);
| leijia = cumsum(W); count = 1;
| for i=1:g
| | if count<=g
| | | for j=1:g
| | | | if leijia(j)>randi([0,leijia(end)])
| | | | | E{1,count} = D{1,j};
| | | | | count = count+1;
| | | | | break;
| | | | end
| | | end
| | else
| | | break;
| | end
| end
| D = E; F = randi([1,g],1,0.3*g);
| for i=1:0.3*g
| | for j=1:0.3*m
| | | ind1 = randi([1,m]); ind2 = randi([1,m]);
| | | ind3 = randi([1,n]); ind4 = randi([1,n]);
| | | [D{1,F(i)}(ind1,ind3), D{1,F(i)}...
| | | (ind1,ind4)] = deal(D{1,F(i)} (ind1,ind4),...
| | | D{1,F(i)}(ind1,ind3));
| | | [D{1,F(i)}(ind2,ind3),...
| | | D{1,F(i)} (ind2,ind4)] = deal(D{1,F(i)}...
| | | (ind2,ind4),D{1,F(i)}(ind2,ind3));
| | | [D{1,F(i)}(ind1,:),D{1,F(i)}(ind2,:)] =...
| | | deal(D{1,F(i)}(ind2,:),D{1,F(i)} (ind1,:));
| | end
| end
end
ZYZ = find(W == min(W),1); ZYJ = JL{1,ZYZ};
plot(ZD,'-.g','linewidth',1.5), hold on
plot(ZX,'-r','linewidth',1.5),plot(JZ,':k','linewidth',1.5),
plot(ZS,'--b','linewidth',1.5)
```

By running the above program, the calculated results are shown in Figure 4. The optimal solution of the previous example is found at the 99th iteration, and the optimal value of the corresponding fitness function is 7270, which is the sum of the moving shelf position distance and the waiting cost involved in the example problem. The total time cost required for the batch of orders can be derived from the time required for each position of the shelf to move.

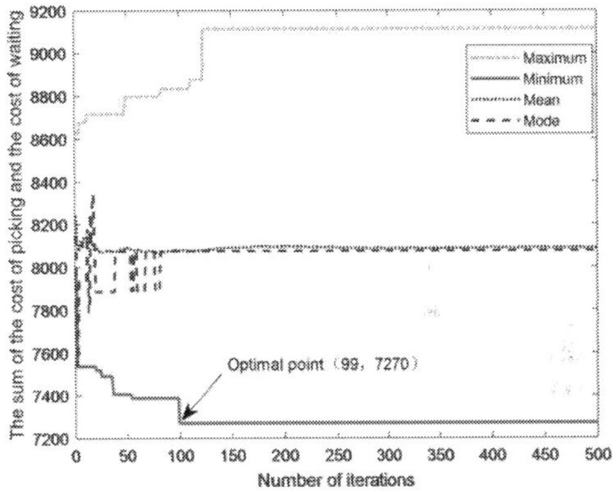

Figure 4 calculation results

Starting from the fitness function to achieve the optimal calculation results, the optimal picking order of 50 orders in this example can be further traced, as shown in Table 6, which is an optimal solution for the research problem.

Table 6 results of picking order

Picking	Order	Picking	Order	Picking	Order
1	O29	18	O9	35	O2
2	O31	19	O28	36	O16
3	O24	20	O50	37	O23
4	O14	21	O44	38	O7
5	O38	22	O46	39	O13
6	O33	23	O35	40	O42
7	O15	24	O32	41	O21
8	O1	25	O20	42	O11
9	O12	26	O25	43	O45
10	O19	27	O22	44	O10

11	O8	28	O43	45	O17
12	O18	29	O5	46	O27
13	O4	30	O48	47	O6
14	O49	31	O40	48	O41
15	O39	32	O26	49	O30
16	O36	33	O37	50	O34
17	O3	34	O47		

In addition, through further analysis of the output results of Matlab, if opening the shelves with a random sequencing, the time cost could be very high. Even if only concern the lowest channel opening time without optiming the order selection sequencing, the mean and maximum value of the fitness will reach 8100 and 9150 respectively. The model constructed in this paper can not only greatly reduce the time cost of picking activities in a completely random state, but also use the genetic algorithm to further optimize the order picking sequencing based on the optimization of each order picked. The time cost of the whole batch of orders is further reduced by 10~25%. In addition, as the scale of intensive mobile shelves increases and the number of orders increases, the optimization effect will be more significant.

7. Conclusions and prospects

In summary, with the application of intensive mobile shelves, while improving the utilization of storage space, it also puts forward higher requirements for order picking activities. Based on the analysis of the application and research status of intensive mobile shelves, this paper studies the selection of bulk orders under single-channel intensive mobile shelves from the following aspects:

① Under the single-channel intensive mobile shelf, the sorting problem of single order is transfer into TSP ;

②Construct a two-stage model and take the minimum total cost of order picking and waiting as the objective in a single period;

③Based on the basic idea of genetic algorithm, the algorithm is designed in detail and implemented by Matlab programming.

④The validity of the model and the solution method are further proved by an example.

Future research can be expanded from the following aspects:

① Concern the mathematical model, there may be more than one algorithm and more than one optimal solution for this kind of problem. How to improve the algorithm and find these optimal solutions more quickly and effectively can become a further study topic;

② This problem is a NP hard problem. The complexity of the problem will increase exponentially with the increase of the scale of the intensive mobile shelves and the number of orders. For the problems with a very large scale, we need to find more suitable algorithm;

③ the assumptions impact a lot on the model. It is necessary to make more adjustments and improvements of the model to fit for different conditions.

References

1. Wang Cong,Jin Wei,Ma Lin,Wang Wei.Sorting optimization of outsourcing orders for auto parts considering scheduling constraints[J].Industrial Engineering and Management,2016,21(04):106-113.
2. Wu Yingying, Meng Xiangxu, Wang Yanyan, Hu Jinchang. Order Sorting Optimization of "Goods to People" Picking System[J]. Journal of Mechanical Engineering, 2016, 52(04): 206-212.
3. Yu Yangyang. Optimization method of picking path based on genetic algorithm[J]. China Science and Technology Information, 2019(08): 91-94.
4. Shao Zezhen, Dong Baoli. Multi-variety and small-volume order batching method based on improved genetic K-means algorithm[J].Journal of Zhejiang Sci-Tech University(Natural Science), 2018,39(06):732-738.
5. LI Yongwei,LIU Shu\'an,GUO Jinqin.Study on the Optimization of Warehouse Picking Path Based on Double-layer Genetic Algorithm[J].Journal of Taiyuan University(Natural Science),2018,36(02):46-51.
6. Pan Chenghao, Guo Min. Simulation of Batch Picking Path Planning for Warehouse Logistics Robots[J]. Computer and Modernization, 2017(02): 12-16.
7. Yu Yangyang. Optimization Method of Picking Path Based on Genetic Algorithm[J]. China Science and Technology Information, 2019(08): 91-94.
8. Shao Zezhen, Dong Baoli. Multi-variety and small batch order batching method based on improved genetic K-means algorithm[J].Journal of Zhejiang Sci-Tech University(Natural Science), 2018,39(06):732-738.
9. Li Yongwei, Liu Shuan, Guo Jinqin. Research on Warehouse Picking Path Optimization Based on Double-layer Genetic Algorithm[J]. Journal of Taiyuan University: Natural Science Edition, 2018, 36(02): 46-51.
10. Pan Chenghao, Guo Min. Simulation of Batch Picking Path Planning for Warehouse Logistics Robots[J]. Computer and Modernization, 2017(02): 12-16.
11. He Yi. Optimization of order picking path in logistics distribution center under genetic algorithm[J].Commercial and Economic Research,2016(21):110-111.
12. Xia Li, Hou Shiwang, Liu Jun. Research on multi-lane picking operation scheduling of single stacker based on hybrid algorithm[J]. Industrial Engineering, 2016, 19(05): 33-38.

Assessment and Characterizing Mechanical Damage in Packaged Bananas in the Post-harvest Supply Chain

Fernando I.[1, 2, a], Fei J. [1,2], Stanley R.[2,3] and Enshaei H[1,2]

[1]*National Centre for Ports and Shipping, Australian Maritime College, University of Tasmania, Australia.*
[2]*ARC Training Centre for Innovative Horticultural Products, University of Tasmania, Australia.*
[3]*Centre for Food Innovation, Tasmanian Institute of Agriculture (TIA), University of Tasmania, Australia*

Abstract. Quality deterioration in packaged bananas caused by mechanical damage along the post-harvest supply chains (SC) remains obscure until the packages are unpacked for sale at the retail stores and therefore, the mechanism of damage occurrence remains unclear. This study assessed the mechanical damage levels of bananas in 300 cartons from pack houses to retail stores in Australia. The damage across the SC were shown to be progressive across the SC. Bruising and neck damage levels in bananas increased from the distribution centre (DC) to the retail stores. Mechanical damage in unripe bananas was influenced by the package location in the stacked-pallet, and the transport and handling of packages within the last-mile of the SC (DC to retail) further exacerbated the damage. This study further characterized the damage development by subjecting packaged bananas to simulated vibration, top-load compression and drop impact. It was revealed that the exposure to vibration resulted in rubbing; top-load package compression contributed to bruising and, the drop impact caused severe neck injuries in bananas. The knowledge of the damage incidences across the SC and causes of damage may contribute to the development of interventions targeted at improving the quality of bananas in the post-harvest SC in Australia.

1 Introduction

In most post-harvest supply chains (SC), a certain degree of mechanical damage to fresh produce during the distribution is unavoidable. Appearance quality of fruits is a major determinant for the consumer's willingness to purchase [1]. Previous studies have shown that cosmetic damages in bananas resulted in poor consumer acceptance and marketability [2]. Bananas have been crowned as the most sold supermarket produce in Australia [3] and their quality is of a great interest not only to the banana industry but also to the major supermarket chains which account for over 74% of the total banana sales [4] in Australia. Bananas, like any other fruit are subjected to mechanical stresses from harvest through to consumers which is one of the major factors for the post-harvest quality deterioration [5-7]. It was estimated that 5-8% of the gross volume of bananas has been accounted as wastage in the post-harvest SC in Australia which amounts to AU$ 46-73 million per annum [8]. Therefore, mechanical damages in bananas have been a significant concern for both the banana and retail industries in Australia, requiring a systematic investigation.

Like other fresh produce, bananas also exhibit a degree of physio-chemical changes due to mechanical injury. One of the most significant and immediate consequence of mechanical damage is the attack of microorganisms, which may cause further decay in fruits [9]. Mechanical damages cause browning in bananas due to the oxidation of poly phenols [10]. Maia, et al. [9] found that mechanical damage caused higher enzyme activity in the affected area followed by alternations in colour, flavour, rapid tissue softening and induced ripening. It was also showed that mechanical damage induced respiration levels in bananas [7, 10]. Chukwu, et al. [11] further reported that abrasion damage caused rapid moisture and weight loss compared to impact bruising in bananas. Similarly Maia, et al. [10] reported that abrasion and cut damages resulted in the highest fresh weight loss in bananas compared to the other treatments. Furthermore, Maia, et al. [10] showed that mechanical damage expedited ripening and influenced the rate of conversion of starch into total soluble sugars (TSS) in the pulp. These findings conclude that mechanical injury induce ripening, hence reduce the effective shelf-life of the fruit. The most significant consequence of these alterations would be the implications on fruit quality and the perceived value of bananas, which can directly affect the marketability.

Mechanical damage in fruits can occur due to compression, impact and vibration when fruits are transported in packages or in bulk bins [12-14]. Bruising can occur due to impacts or compression forces [15, 16] resulting in peel and/or flesh damage in bananas

[a] Corresponding author: Indika.fernando@utas.edu.au

depending on the severity of the impact/pressure. Damage caused by friction occurs due to the relative movement of the fruit against other contacting surfaces. Abrasion damages such as rubbing in bananas can occur due to vibration or when bananas are handled in clusters (e.g. when clusters are rubbing against conveyer belts in a packing line). Several studies showed that vibration during transport was a critical factor for the development of mechanical damage in bananas [6, 17, 18] and prolonged exposure to vibration could escalate the incidence of damage [5]. Studies on other packaged fruits revealed that the vibration damage can be severe in loosely packed fruits compared to the tightly packed fruits [19, 20]. Unlike most other fruits, bananas may also exhibit stem end tearing or neck damage. Neck injuries are caused by separation of stem from the fruit [21] which can affect the quality and the consumer acceptance at the retail stores [2]. However, limited knowledge exists on the exact mechanism of the occurrence of neck injuries especially when banana clusters are packed inside corrugated paperboard cartons during distribution.

Studies assessing the occurrence of mechanical damage in bananas across the SC, from the farm gate to the retail-shelf are limited. Maia, et al. [22] investigated mechanical damage at four distinct locations in the banana SC (pre-selection, after packing, after transport and retail market) and concluded that the damaged levels increased and accumulated along the chain. A similar study revealed that the damage to bananas was progressive along the SC and the highest damage occurred during the post-harvest distribution [23]. However, damages caused by different mechanical stresses in bananas are not well defined in literature and a deep understanding of different types of damages is necessary to determine the causes of such damage. Previous studies revealed several types of cosmetic damages in bananas (Fig.1) [2] however their mechanism of occurrence, especially within a package and during the handling across SC has not been well understood. Therefore, this study aims to (1) assess and quantify the mechanical damages in bananas across the post-harvest SC in Australia and (2) characterize different types of damages in packaged bananas caused by induced mechanical stresses including vibration, compression and transient impacts.

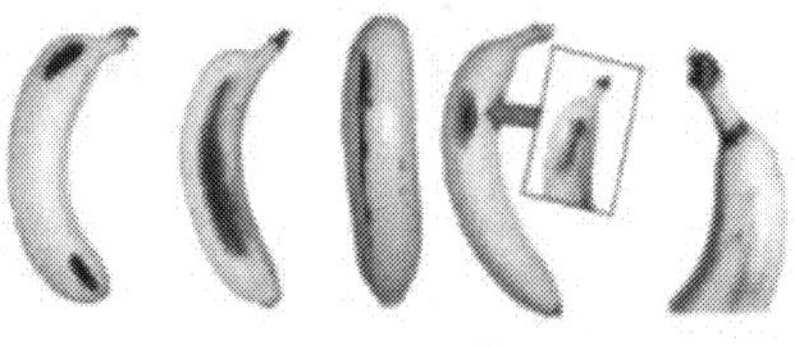

Fig 1: Different Types of Cosmetic Damages in Bananas [2]

2 Materials and Methods

2.1. Damage Assessment in Packaged Bananas

To understand the occurrence of cosmetic damages in bananas, the damage levels were assessed in three locations across the chain, from the pack-houses through the distribution centre (DC) to the retail stores (Fig.2). Bananas are commonly packed in three grades (i.e. premium, extra-large and large) for different markets and the 'premium' grade (selected and packed as the highest grade) is intended for the major supermarket chains in Australia. Cartons packed with 'premium' grade bananas were selected for this study because they are generally graded before the packing process with minimal pre-existing damage. Having a minimal (or none) existing damage in bananas at the initial stage of the SC (i.e. pack-house) is essential for the damage comparison to distinguish the damage development at the later stages of the SC (i.e. DC and retail stores).

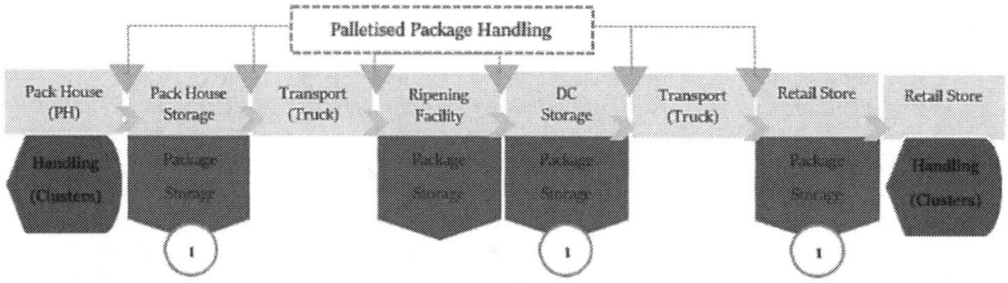

Figure 2: Simplified Post-Harvest Banana Supply Chain [(I) – Damage inspection points]

Five SCs were selected for the study, originating from five pack houses in Northern Queensland to the retail stores in Melbourne (over 3000 km via the road route) through a regional DC in Melbourne. A sample of 100 packages were selected randomly in each location (i.e. pack-house, DC, retail store) for damage assessment. Twenty packages were assessed in each pack house and retail store. The stacked-tier of the selected packages in the pallets in each location is given in Table 1. Banana packages were stacked in 10 tiers on wooden pallets during the distribution from the pack-houses to DC and stacked only up to five tiers (mostly consolidated with other produce packages) from the DC to retail stores. Therefore, number of cartons inspected in each tier was 10 and 20 in the DC and retail stores respectively. A total of 300 cartons packed with 3,240 clusters of bananas (15,261 individual fingers) were assessed for mechanical damage and quality defects across the SC.

Table 1: Number of Packages Inspected in Each Tier of the Pallet in the DC and Retail during Random Inspection

Location	No. of Packages Inspected in Each Tier of the Pallet										Total
Tier Number	1	2	3	4	5	6	7	8	9	10	
Pack-house	10	10	10	10	10	10	10	10	10	10	100
DC	10	10	10	10	10	10	10	10	10	10	100
Retail	20	20	20	20	20	-	-	-	-	-	100

2.2. Characterizing Mechanical Damage in Packaged Bananas

Packaged bananas were subjected to simulated vibration, top-load compression and simulated handling (drop impact) to determine the causes of different types of mechanical damage in bananas. Forty (40) sample banana cartons were arranged for the experiments. All sample cartons were sourced from the same batch of fruits, harvested/packed from the same farm/pack-house in Tully, Queensland. Fifteen (15) cartons were packed with unripe bananas (stage two) and the rest were packed with ripe bananas (stage six) as per the ripeness chart in Ekman, et al. [2]. The cartons were packed in three-layers (5-7 clusters per layer arrangement) inside each package. All sample cartons were pre-inspected for existing damage and the clusters with pre-existing damages were replaced with non-damaged clusters prior to each experiment. All cartons with unripe bananas were conditioned for 48 hours at 13.5 °C and 50-60% relative humidity (RH) and the ripe bananas were stored in 15-18 °C at 90-100% RH (inside the ripening chamber) for 48 hours before the experiments.

2.2.1 Simulation of Package Handling and Top-load Compression

A drop-testing machine (Lansmont PDT 80, CA, USA) was used to drop packaged bananas (ripe) from heights of 0.3 m and 0.5 m (Fig. 3). Each carton was dropped from these specified heights three times during each test and the experiment was conducted in a randomized block design in pentaplicates (5 x 2). For the package compression experiment, ten cartons of ripe bananas were subjected to top-load compression by a compression testing machine (Instron 1185, MA, USA). Five sample cartons with ripe bananas were crushed to 16 mm and the other five cartons were crushed to 25 mm vertical displacement levels. Each package compression experiment was conducted in randomized block design in pentaplicates. All the sample packages were stored in 13.5 °C and 50-60% RH and the damage levels in packaged bananas were assessed 12 hours after each experiment by the procedure described in section 2.3 to derive a damage score for each package.

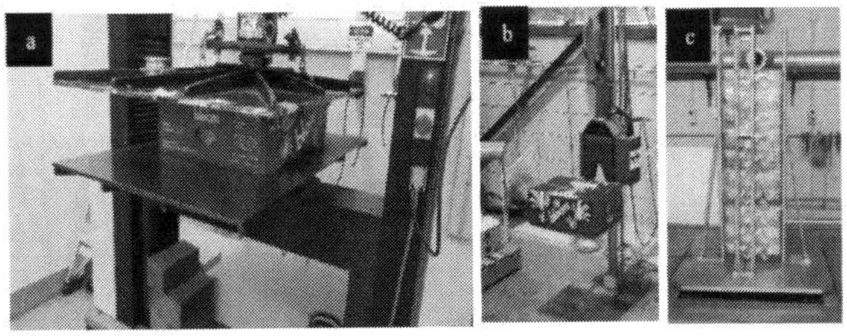

Figure 3: Characterizing Mechanical Damage by (a) Top-load Compression, (b) Drop Impact and (c) Simulated Vibration

2.2.2. Simulated Vibration Testing

A single vertical column of ten banana cartons (unripe) was subjected to simulated vibration intensity of 0.36 g for a duration of 3 hours by an electro-hydraulic vibration simulation table. The vibration profile of ASTM-D4169 was used[24]. The column was triple-stretch-wrapped with transparent stretch film and supported by vertical steel columns fixtures, attached to the table from either side to avoid the collapsing of the packages during the test (Fig.3). After each experiment the cartons in the bottom (1st tier), middle (5th tier) and top (10th tier) were removed from the column for damage assessment and the three sample positions were stacked with fresh (non-vibrated) cartons. The packages in the rest of the positions of the column were reused as dummy packages in each repeat experiment. The damage level in bananas stacked in the three sample positions of the column were evaluated 12 hours after each experiment by the procedure described in the section 2.3. This experimental procedure was repeated for ripe bananas. All vibration simulation experiments were conducted in pentaplicates in a randomized block design (3 x 5).

2.3. Damage Assessment and Analysis

The damage in packaged bananas was assessed by a 0-4 point hedonic scale (0- No damage; 1- Trace damage; 2- Slight Damage; 3- Moderate Damage; 4- Severe Damage) in accordance with the visual quality assessment chart [2]. A visual damage index (VDI) score was calculated by the Eq.1 for each package similar to the equivalent bruise index for apples [25]. The mean VDI score was used for the comparison of damage levels in bananas occurred after each treatment.

VDI (%) = (0.1 x Trace Damages (%) + 0.2 x Slight Damages (%) + 0.7 x Moderate Damages (%) + 1.0 x Severe Damages (%) (Eq.1)

One-way analysis of variance (ANOVA) and Turkey's Honest Significant Difference (HSD) test were performed to test for significant differences within the means by GraphPad® Prism 7 statistical analysis package (GraphPad Software, La Jolla, CA, USA).

3. Results

3.1. Mechanical Damage along the Supply Chain

Different types of mechanical damages exhibited in different stages of the SC during the damage inspection is given in Fig. 4. The levels of neck damage, bruising and various abrasion damages (i.e. scuffing, fruit rub, and blackened rub) by the stacked-tier in the pallet as revealed in the DC and retail stores are given in Figure 5 and Figure 6 respectively. Damage assessment in packaged bananas revealed that the accumulated damage level was escalated by 1.8% to 8.9% from the pack houses to the retail stores (Fig. 4). The highest increment in damage (5.1%) was revealed in abrasion type damage while bruising and neck damage exhibited a relatively less increment (<2%) from the pack houses to the retail stores. The highest increment of damage between two SC nodes was revealed in abrasion damage between the pack houses and DC (2.6%) and the abrasion damage level has further worsened between the DC and retail stores (2.5%). The level of bruise and neck damage at the pack houses marginally increased at the DC however neck damage levels had nearly doubled between the DC and retail stores.

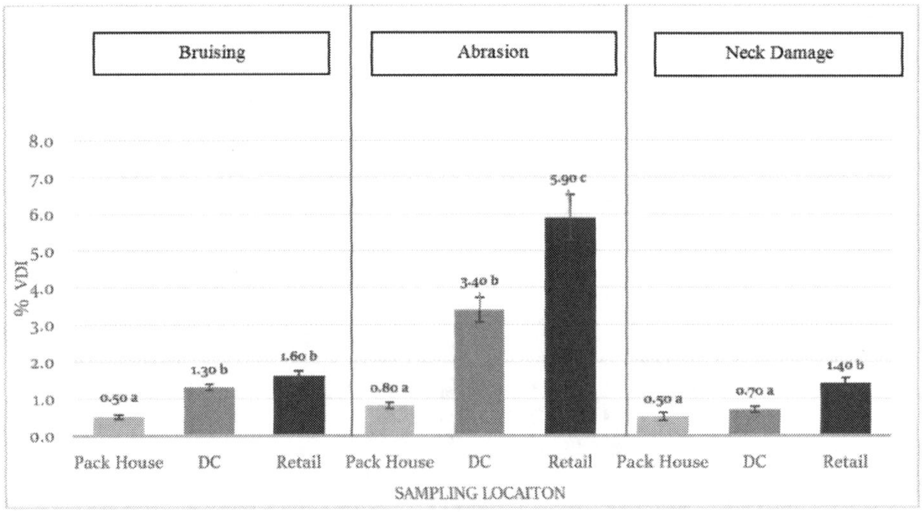

Figure 4: Mechanical Damages in Bananas by the Sampling Location in the Supply Chain [For each damage Category Different Characters Indicates Significantly (P<0.01) Different Results]

Figure 5 illustrates the different types of damage in bananas by the stacked-tier which were revealed during the damage assessment at the DC. The best-fitted line (3rd order polynomial) for the overall damage level from tier one to ten elaborates that the highest damage level in a stacked pallet was found in the most top and the bottom packages. The accumulated damage level is minimal in the mid-tier packages of the pallet. Abrasion damage largely contributed to the escalation of the overall damage level in the two bottom tiers and the three top tiers. It is also revealed that the bruise damage and neck injury levels decreased from the bottom to top of the pallet. Statistical analysis revealed that the bruise damage levels in the first six tiers are significantly different (P<0.01) to that of the top four tiers. Similarly, the neck damage levels of the top four tiers are significantly different

https://doi.org/10.1051/matecconf/201929602004

(P<0.01) to that of the rest of the lower tiers in the pallet. The abrasion damage level in the most bottom tier is significantly different (P<0.01) to the rest of the tiers except for the damage level in the ninth tier. There was no statistical difference in abrasion damage levels from the third tier to the seventh tier however, the highest abrasion damage level was revealed in the top tier packages which was statistically different (P<0.01) to the rest of the tiers. Most of the abrasion damage levels were influenced by the frequent occurrence of fruit rubbing and scuffing especially in the most bottom and top tiers of the pallet.

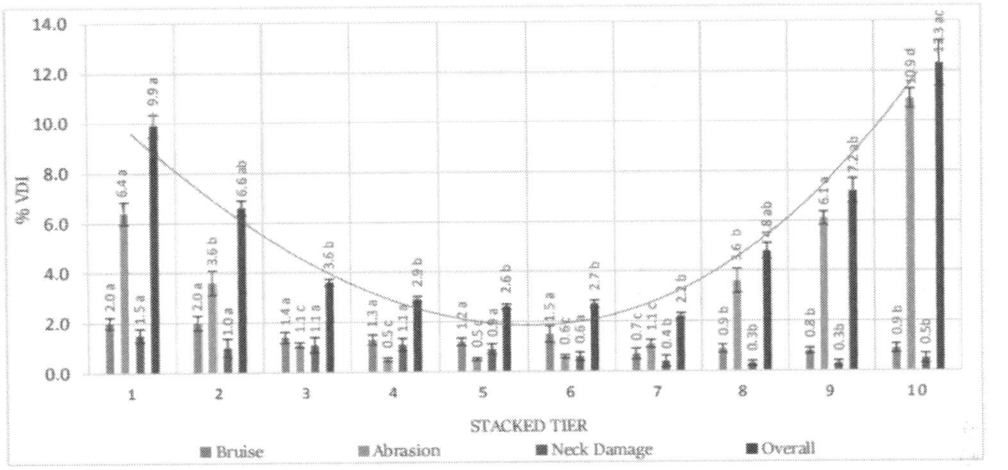

Figure 5: Mechanical Damages in Bananas by the Stacked Tier at the DC [For each damage Category Different Characters Indicates Significantly (P<0.01) Different Results]

As exhibited in Fig. 6 there are no statistically significant differences found for any damage type or for the overall damage level based on the stacked tier of the packages at the retail stores. It was also revealed that, at the retail stores the most critical damage was abrasion. As illustrated in Fig. 7, the most frequent abrasion damage at the retail stores was rubbing damage (70%) including the fruit rub and blacked rub however, the other abrasion damages such as scuffing (17%) and scars (13%) were marginally exhibited.

Inside each package, bananas are packed in three layers (Fig. 8) and the damage level by the layer of fruit inside the carton is given in Fig. 9. It was revealed that a higher incidence of bruising (1.2%) existed in the first layer of fruits inside the carton and the frequency of abrasion type damages such as fruit rub and scuffing was higher in the second and third layer inside the carton. Neck injuries were also escalated in the third layer and the second layer (1.3%) than the first layer (0.1%).

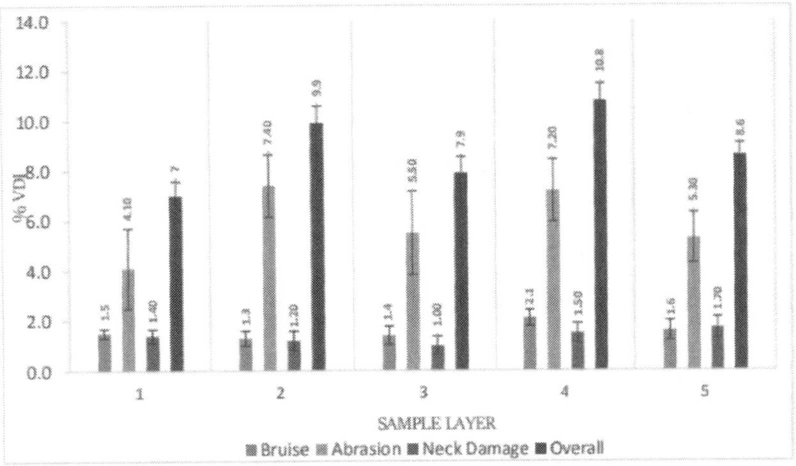

Figure 6: Overall Mechanical Damages in Bananas recorded during Random Inspection at each Stage of SC [No statistically different results (P<0.01) for each damage category have been exhibited]

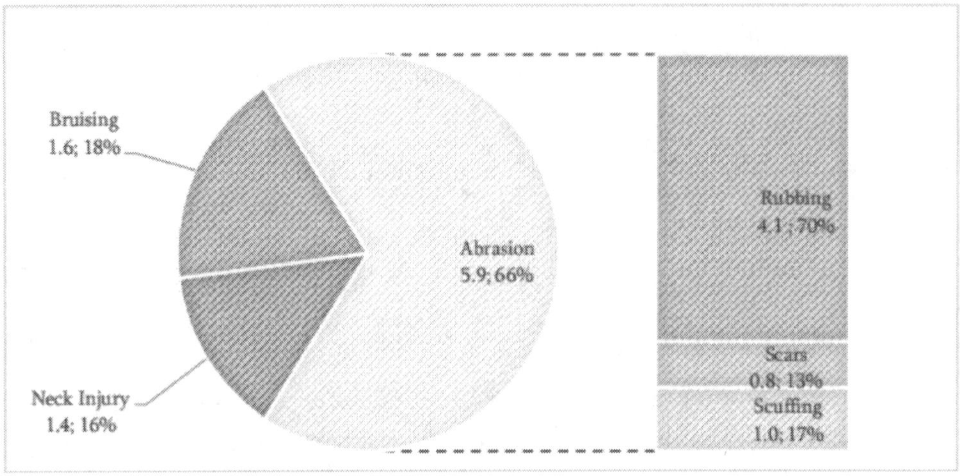

Figure 7: Breakdown of VDI Score and the Percentage (%) by Each Damage Type for the Retail Stores

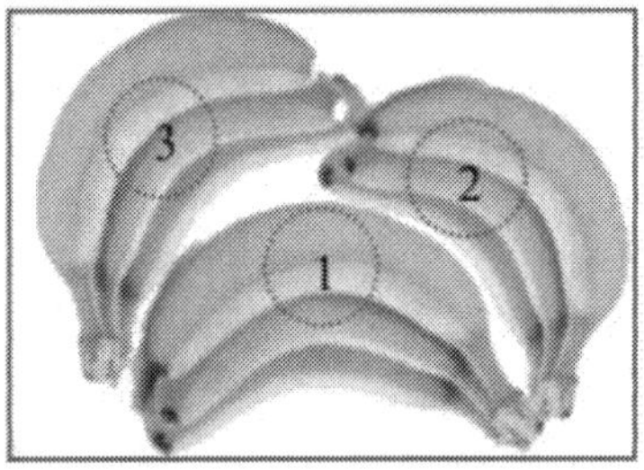

Figure 8: Spatial arrangement of bananas inside the corrugated box packed in three layers (Layer 1, 2 and 3)

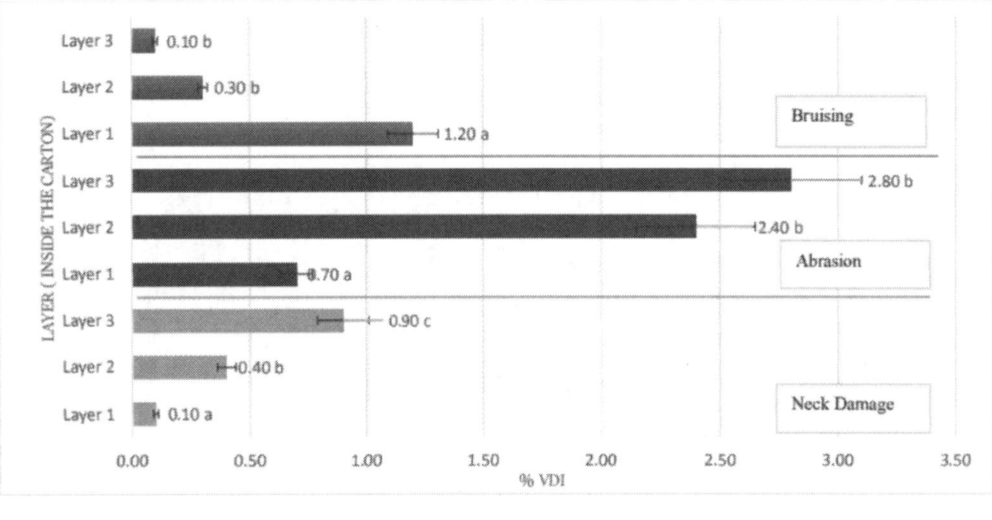

Figure 9: VDI for the Layer of Bananas inside the Carton (Retail) [Different characters (a, b and c) denotes statistically different ($P<0.01$) results for each damage category]

3.2. Characterizing Mechanical Damage in Packaged Bananas

Different types of mechanical damages occurred in packaged bananas due to package drop, top-load compression and vibration (Fig. 10). Package drop mainly resulted in neck injuries and package compression resulted in bruise damage (Fig. 10). In addition to neck injuries, package drop occasionally resulted in minor levels of bruising in ripe bananas. Exposure of packages to simulated vibration mainly resulted in abrasion damage in bananas (Fig. 11). For the same vibration treatment, the damage levels were significantly higher ($P<0.05$) in ripe bananas compared to the unripe (green) bananas (Fig. 11) and majority of the vibration damage occurred in the top two layers of bananas inside the package.

As illustrated in Fig. 12, the neck damage level (VDI %) worsened ($P=0.076$) with the increasing drop height of the packages. Nearly 94% and 91% of the neck injuries (attributed to 0.3m and 0.5m drop heights respectively) were concentrated in the top two layers of bananas inside the package (Fig.12). Bruise damage level (VDI %) in bananas significantly ($P<0.05$) proliferated due to the increased top-load compression of the packages from 16 mm 25 mm (Fig. 12). Nearly 72% of bruising damage occurred under the 16 mm compression and 84% under the 25 mm compression was localized in the first layer inside the package.

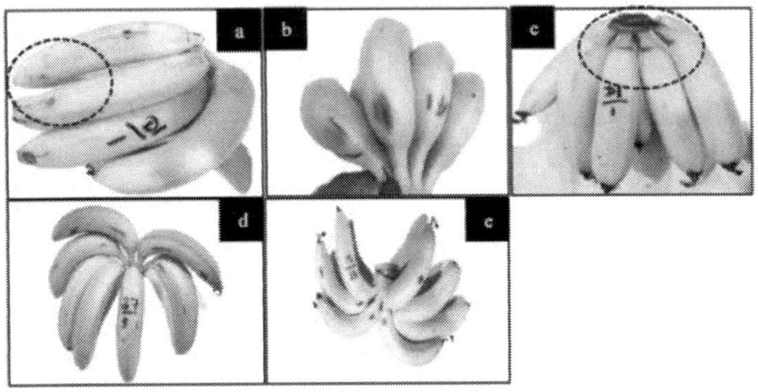

Figure 10: Different types mechanical damages occurred in bananas:
(a) minor (trace) levels of bruising occurred by the package drop impact
(b) shoulder bruising in bananas occurred by top-load compression (25 mm) (c) severe neck damage occurred due to package drop impact
(d) fruit-rub damage in unripe (green) bananas caused by simulated vibration and
(e) fruit-rub damage in ripe bananas

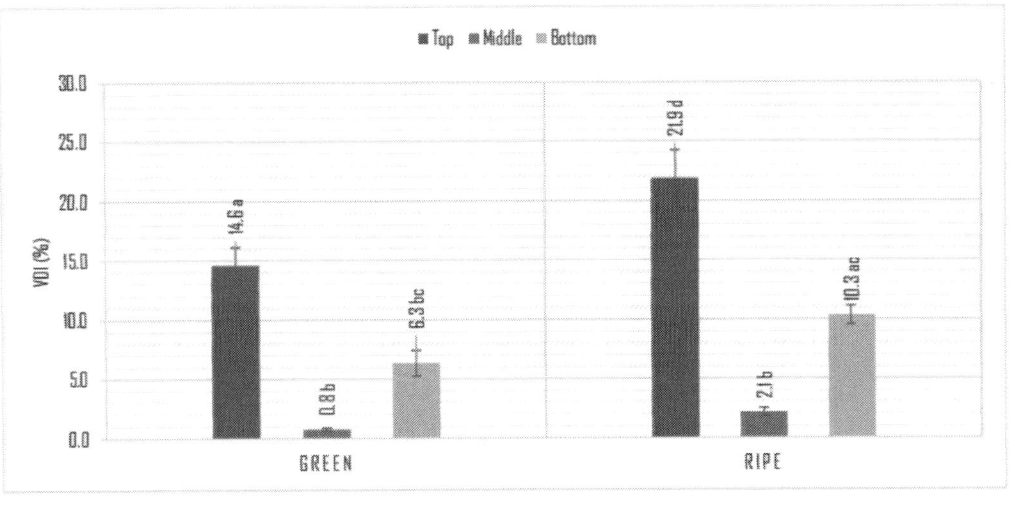

Figure 11: Mechanical damage levels (i.e. abrasion) in green and ripe bananas caused by simulated vibration [Different letters indicate significantly different results ($P<0.05$)]

 https://doi.org/10.1051/matecconf/201929602004

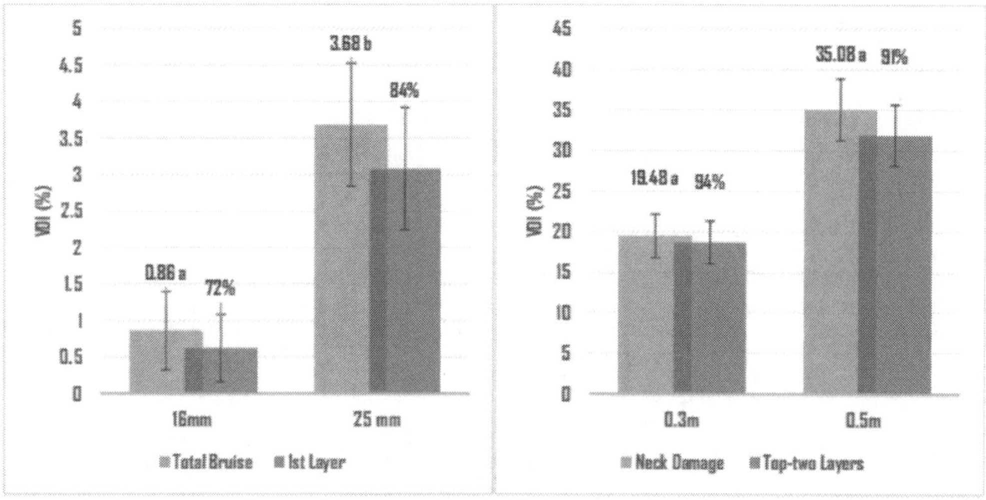

Figure 12: Mechanical damage levels occurred during the top-load compression (left) and drop-impact tests (right) [Different letters indicate significantly different results (P<0.05) in each test]

4. Discussion

The results from the damage assessment indicates that all types of damages in bananas (i.e. bruising, neck injuries, abrasion) exacerbated along the SC, resulting in poor visual quality at the retail stores. At all three stages of the SC, abrasion damages were the most frequently reported. It was also revealed that most fruit rub damages occurred between pack houses and DC while blackened rub and scuffing damages were mostly caused from DC to retail stores. These results indicate that damages in bananas are accumulating along the SC, leading to an overall quality deterioration at the retail end. In addition, abrasion damage is the predominant type of damage in post-harvest banana SC. Similar results were reported in previous studies which reported that mechanical damages in bananas were progressive along the SC [9, 10, 22, 23] and the abrasion type damages were predominant in packaged bananas [2].

The simulation tests revealed that ripe bananas were more susceptible to vibration damage compared to the unripe (green) bananas (Fig. 11). This could be the reason for the disproportionate increment of damage during the last-mile distribution of ripe bananas (from DC to retail). The results show that the interstate transport and handling of bananas (over several thousand kilometers) resulted in 2.6% damage increment in unripe (green) bananas while the distribution of ripe bananas to retail stores located less than 100 km from the DC caused 2.5% increment in abrasion damage. Similarly neck injuries in bananas nearly doubled between the DC to retail stores.

The drop tests confirmed that one of the major reasons for neck injuries in bananas is the manual handling of packages. The significant increase in neck injuries in the latter part of the SC (from DC to retail) indicates that softer ripe bananas become more susceptible to neck damage than the harder unripe (green) bananas during handling. Therefore, it is essential that packages with ripe bananas are handled cautiously, especially during the palletizing and package consolidation i.e. staking different produce packages together in mix-pallets at the DC to be dispatched to different retail stores to minimize neck injuries. Similar results were reported in the previous studies inciting that the susceptibility of bruising in bananas escalated with ripening [10, 26] and thus, carefully handling and storage packages will be integral to improving the quality.

Most of the abrasion damage recorded at the DC was concentrated in the top and bottom tiers of the pallet (Fig.5), while the middle tier packages (tier 4 to tier 7) exhibited a minimal level of abrasion (i.e. fruit rub). Most of the fruit rub damage in the bottom and top tiers of the pallet were caused by relative motion of fruits in the package, which was associated with the transmissibility of vibration in a pallet. The simulated vibration tests revealed similar results, confirming that more vibration damage occurred in the bottom and top tier packages with minimal damage occurring in the middle tiers (Fig. 11). This result coincides with the previous studies which have shown that the highest vibration damage in fruits were often revealed in the most top tier packages of a pallet\column [27, 28]. However, the reason for reduced damage in the middle tiers of the pallet requires further investigation, including examining the vibration dampening properties of corrugated packaging.

Bruising and neck damage levels escalated in the bottom tiers of the pallet, revealing the highest levels of damage in the most bottom tier packages (Fig.5). Top-load compression tests confirmed that bruising in bananas occurred mostly due to the compression of the packages from the upper tiers. The weight from the upper tiers act on bananas as quasi-static compression during storage, handling and transport. Dynamic compression forces on packages during transport may also occur due to shocks encountered by palletized packages. However, more research is necessary to determine the influence of transient shocks during transport on the occurrence of damage in packaged bananas. Despite the relatively lower bruise and neck damage levels in the upper tiers of the pallet, the overall accumulated damage levels still increased in the top tiers, due to the high levels of abrasion damage (Fig.5).

The presence of increased neck injuries in the bottom tier packages at the DC indicates that neck injuries are not only caused by package handling but also associated with the weight of the fruits stacked on-top of a given package. It was revealed during the damage assessment that occasional package failure in the bottom tiers was closely associated with a higher incidence of neck injuries in bananas. However, the reasons for such sporadic package failure was not always straightforward but observed to be associated with misaligned stacking of packages on pallets and moisture absorption by paperboard cartons during the ripening process.

Most abrasion damage was concentrated in the top two layers inside the package (Fig.9). This can be due to the increased freedom of movement of fruits when exposed to vibration and the movement or bouncing of clusters inside the package due to the shock impacts (e.g. package dropping). When individual fingers are free to move inside the package, it results in rubbing between the fingers and clusters. The weight of the two upper layers and the inter-locking mechanism of clusters inside the package (Fig.8) restricted the freedom of movement for clusters in the first layer of the carton resulting in reduced abrasion damage in the first layer. During the package drop tests, the energy of the transient shock events (e.g. by hitting the floor/ ground) was partially transmitted to banana clusters packed inside. The top two layers of fruit (especially the most top layer) had more freedom to bounce resulting high incidence of neck injuries in the top two layers inside the package. Similar results were reported by Ekman, et al. [2].

4 Conclusion

Minimizing quality deterioration in bananas requires knowledge on the incidence of mechanical damage along the SC and the mechanism of damage occurrence in packaged bananas. This study evaluated the cosmetic damage levels in bananas along the post-harvest SC and experimentally tested and confirmed the causes for damage occurrences. The damage levels were found progressive along the SC from the pack houses to the retail stores. The damage levels in green bananas were significantly influenced by the package position in the stacked-pallet. The abrasion damage was mostly concentrated in the top and bottom tier packages while both neck injuries and bruise damage were frequently found in the middle and lower tier packages of the pallet. All types of damages further proliferated during the distribution of ripe bananas to the retail stores and bruising and neck injury levels nearly doubled from DC to retail stores. The laboratory simulation experiments confirmed the observed field damages and revealed that top-load compression on the packages resulted in bruise damage in bananas while the simulated package drop resulted in severe neck injuries. Exposure to simulated vibration contributed to the development of fruit rub damage in the most top and bottom tier packages of the stacked column however, the middle tiers exhibited minor level of rubbing damage. Minimizing bruising and neck injuries in bananas requires careful storage and handling of packages especially during the distribution from the DC to retail stores. Further research is warranted to examine the failure mechanisms of paperboard packaging caused by moisture absorption during the ripening process as partial package failure was associated with increased neck injuries in bananas. Minimizing mechanical damage caused by vibration also requires further investigation on the vibration transmissibility and dampening properties of pallets and packaging.

References

1. S. R. Jaeger, L. Antúnez, G. Ares, J. W. Johnston, M. Hall, and F. R. Harker, "Consumers' visual attention to fruit defects and disorders: A case study with apple images," *Postharvest Biology and Technology*, vol. 116, pp. 36-44, 2016.
2. J. Ekman, N. King, S. Lindsay, and K. Gething, "Improved handling in banana supply chains," 2011.
3. J. Margetts, "Australian Banana Industry- Strategic Investment Plan," Plant & Food Research Australia Pty Ltd 2014.
4. C. Day. (2018, 14/09/2018). Despite Higher Prices, Aussies Still Love their Bananas. Available: https://www.nielsen.com/au/en/insights/news/2018/despite-higher-prices-aussies-still-love-their-bananas.html
5. B. K. Dadzie and J. E. Orchard, *Routine post-harvest screening of banana/plantain hybrids: criteria and methods*: *International Plant Genetic Resources Institute (IPGRI)*, 1997.
6. W. Wasala, D. Dharmasena, C. Dissanayake, and B. Tilakarathne, "Feasibility Study on Styrofoam Layer Cushioning for Banana Bulk Transport in a Local Distribution System," *Journal of Biosystems Engineering*, vol. 40, pp. 409-416, 2015.
7. J. S. Del Aguila, L. S. Heiffig-del Aguila, F. F. Sasaki, G. M. Tsumanuma, M. das Graças Ongarelli, M. H. F. Spoto, *et al.*, "Postharvest modifications of mechanically injured bananas," *Revista Iberoamericana de Tecnología Postcosecha, Hermosillo*, vol. 10, pp. 73-85, 2010.
8. T. Kitchener, "Project BA13019 :Carton Management inthe Banana Industry," Kitchener Partners Pty Ltd16th December 2015.
9. L. C. B. Maia, V. M. Maia, L. C. C. Salomão, B. P. A. Salles, and R. F. Pegoraro, "Metabolic alterations in banana induced by mechanical damage," *Unimontes Científica*, vol. 17, pp. 27-34, 2015.
10. V. M. Maia, L. C. C. Salomão, D. L. Siqueira, R. Puschman, V. J. G. Mota Filho, and P. R. Cecon, "Physical and metabolic alterations in" Prata Anã" banana induced by mechanical damage at room temperature," *Scientia Agricola*, vol. 68, pp. 31-36, 2011.

11. U. Chukwu, A. Olorunda, and R. Ferris, "Effect of Postharvest Injury on Weight Loss, Ripening Period, and Pulp Loss on Musa spp. Fruits," in *Postharvest Technology and Commodity Marketing: Proceedings of a Postharvest Conference, 2 [9] Nov to 1 Dec 1995, Accra, Ghana*, 1998, p. 132.
12. I. Fernando, J. Fei, R. Stanley, and H. Enshaei, "Measurement and evaluation of the effect of vibration on fruits in transit—Review," *Packaging Technology and Science*, pp. 1-16, 2018.
13. U. L. Opara and P. B. Pathare, "Bruise damage measurement and analysis of fresh horticultural produce—A review," *Postharvest Biology and Technology*, vol. 91, pp. 9-24, 2014.
14. Z. Li and C. Thomas, "Quantitative evaluation of mechanical damage to fresh fruits," *Trends in Food Science & Technology*, vol. 35, pp. 138-150, 2014.
15. N. H. Banks and M. Joseph, "Factors affecting resistance of banana fruit to compression and impact bruising," *Journal of the Science of Food and Agriculture*, vol. 56, pp. 315-323, 1991.
16. N. H. Banks, C. A. Borton, and M. Joseph, "Compression bruising test for bananas," *Journal of the Science of Food and Agriculture*, vol. 56, pp. 223-226, 1991.
17. W. Wasala, D. Dharmasena, T. Dissanayake, and B. Thilakarathne, "Vibration simulation testing of banana bulk transport packaging systems," *Tropical Agricultural Research*, vol. 26, 2015.
18. P. More, D. Behere, S. Housalmal, S. Khodke, S. Jadhav, and A. Kakade, "Assessment of vibration damage to banana bunches simulated by vibration tester," *International Journal of Tropical Agriculture*, vol. 33, pp. 1419-1422, 2015.
19. N. F. Sommer, H. M. Laude, L. J. Berry, H. F. Madsen, S. C. Hoyt, J. N. Luthin, *et al.*, "Surface discoloration of pears: Compression pads between fruit and carton lid effectively prevent transit bruising of jumble-packed Bartlett pears," *Order*, 1957.
20. D. Slaughter, R. Hinsch, and J. Thompson, "Assessment of vibration injury to Bartlett pears," *Transactions of the ASAE*, vol. 36, pp. 1043-1047, 1993.
21. Mohsenin, *Physical Properties of Plant and Animal Materials: Structure, Physical Characteristics, and Mechanical Properties* vol. 2: Gordon and Breach, 1986.
22. V. M. Maia, L. C. C. Salomão, D. L. Siqueira, R. Puschmann, V. J. G. Mota Filho, and P. R. Cecon, "Types and intensity of mechanical damages on'prata anã'bananas along the commercialization chain," *Revista Brasileira de Fruticultura*, vol. 30, pp. 365-370, 2008.
23. F. B. Da Costa, R. Puschmann, S. I. Moreira, J. I. R. Junior, and F. L. Finger, "Survey of mechanical injury in 'Prata Ana'banana during shipping," *Revista Verde de Agroecologia e Desenvolvimento Sustentável*, vol. 5, pp. 72-78, 2010.
24. ASTM, "Packaging and Shipping Container Standards," West Conshohocken, PA: ASTM International, 2016.
25. K. Vursavus and F. Ozguven, "Determining the effects of vibration parameters and packaging method on mechanical damage in golden delicious apples," *Turkish Journal of Agriculture and Forestry*, vol. 28, pp. 311-320, 2004.
26. Y. Yuwana, "Impact Bruise Susceptabilities of Unripe Banana, Mango and Avacado," *Indonesian Food Nutrition and Progress (IFNP)*, vol. Volume 4,, pp. 56-60, 1997.
27. B. Jarimopas, S. P. Singh, and W. Saengnil, "Measurement and analysis of truck transport vibration levels and damage to packaged tangerines during transit," *Packaging Technology and Science*, vol. 18, pp. 179-188, 2005.
28. D. Fischer, W. Craig, A. Watada, W. Douglas, and B. Ashby, "Simulated in-transit vibration damage to packaged fresh market grapes and strawberries," *Applied Engineering in Agriculture*, vol. 8, pp. 363-366, 1992.

An Approach to Develop the Sustainable Warehousing Assessment Model

Magdalena Malinowska[1,a]

[1] *University of Szczecin, Faculty of Management and Economics of Services, Cukrowa 8, 71-004 Szczecin, Poland*

Abstract. The aim of the article is to present an approach to define the set of criteria applied to create the sustainable warehousing assessment model. The criteria were identified on the basis of expert analysis and the literature review including the scientific approach as well as standards applied for the purpose of assessing the level of sustainable buildings. It allowed to include in the proposed model the broad spectrum of factors, which affect the warehousing sustainability in economic, social and environmental aspect. As a solution utilized for the purpose of sustainable warehousing assessment model creation the COMET method (characteristic object method) was applied. This constitutes the basis for applying the fuzzy set theory to propose the characteristic of each criteria used for the assessment.

1 Introduction

However the problem of sustainable development was firstly discussed 300 years ago by Hans Carl von Carlowitz, the term became popular in 80-ties of the last century [1]. Firstly joined with the ecology, due to conducted research was considered taking into account its interdisciplinary character, manifesting the need for integration in key areas: economy, society and environment [2].

Nowadays, the sustainable development is more often used in logistic due to the fact that it is the human activity present and influencing at various areas of our life. In the mid-2000s, the concept was highly promoted regarding to building and modernizing warehouses and distribution centers towards environmentally-friendly solutions, reducing operating costs and creating more favorable working conditions. However the global slowdown in economic growth slightly weakened this trend, the idea to apply in warehouses the solution, which are increasing the level of sustainability is still the focus of interest for many companies [3]. More and more popular is becoming the certification of warehouses on the basis of existing building standards.

This article also considered the issue of assessing the level of sustainability in the context of warehousing and presents an approach for description of criteria that are taken into account as field of measurement.

2 Literature review

The sustainable development is today widely discussed in various areas of business activities. On the field of logistics, there are many research considered in particular the sustainable supply chain management or sustainable transport system. There are also some studies focusing on the context of sustainable warehousing area to create a unit equipped by appropriate technological and organizational solutions, which allows to operate maintaining the high standards of work, reduce the negative impact on the environment and achieve the lover cost of its functioning [4]. This assumptions are compatible with the considerations of sustainable warehouse given by Żuchowski, which requires "a set of organizational and technological solutions whose aim is to efficiently execute warehouse processes, with the highest social standards met, with the lowest possible environmental impact and taking financial effectiveness into account" [1]. However this definition indicates different dimensions of sustainability concerning the building design, application of advanced technology to support the warehouse processes and application of social policy to ensure higher standards of work, nowadays the problem of sustainability of warehouse is measured taking into account the existing building certification standards: BREEAM (Building Research Establishment Environmental Assessment Methodology), LEED (Leadership in Energy and Environmental Design), HQE (Haute Qualité Environnementale), DGNB (Deutsche Gesellschaft für Nachhaltiges Bauen) [5-7]. This standards require that location, design, construction, maintenance and demolition of the building should provide the effective utilization of energy, water and materials which in consequence contribute to provide benefits for owners, higher social standards, better financial effectiveness and more environmental friendly environment. The most often applied certification systems are LEED and BREEAM.

LEED [6-7] (v4 version) takes into account 9 categories with specified aspects which are the basis of

[a] Corresponding author: magdalena.malinowska1@usz.edu.pl

the assessment: an integrative processes (e.g. the ability to use innovative approaches to design and construct the building, selection of a project team), location and transportation (e.g. neighborhood location, efficient transportation options), sustainable sites (e.g. minimization of pollution, naturally inclusion to ecosystem), water efficiency (e.g. reduction of water consumption, application of innovative technologies to avoid wastage of water), energy and atmosphere (e.g. use of the renewable energy sources, optimization of energy consumption, application the technological solutions to reduce energy consumption), material resources (e.g. utilization of recycled materials, use of certified materials), indoor environmental quality (e.g. ventilation system efficiency, possibility of temperature control, CO2 monitoring), innovation in design (aspects not included in the basic criteria), regional priority (entering the regional strategy). BREEAM [6-7] distinguishes the following categories: management (e.g. overall management policy, site management, procedural issues), energy (e.g. CO2 emission, light energy consumption), health and wellbeing (e.g. amount of daylight in rooms, temperature and air quality, acoustics), materials (e.g. use of materials obtained from legal and local sources with appropriate ecological certificates), transport (e.g. CO2 emission, location-related factors), water (e.g. application the solutions limiting water consumption), waste (e.g. applied waste management policy), pollution (e.g. water and air pollution impact), land use and ecology (green areas management, biodiversity of flora and fauna).

The analysis of given standards lets to notice that the considered aspects affecting the sustainability are convergent in both cases. However they are focused on the construction of buildings and do not relate to specify of warehouse and the realized processes. The criteria like water and energy consumption, air quality, application of environmental friendly materials or land development are required to be the core of prepared warehouse assessment model, but the way they are measure and the given scale must refer to warehouse conditions.

As indicate [4] and [8], the concept of sustainable warehousing goes beyond the building facilities. [4] described eight sustainable warehousing constructs within which there are included elements presenting various dimensions of sustainability. It allows to propose best practice activities, that can be used as a roadmap for setting future targets and goals towards achieving higher level of sustainable development.

The above remarks were the basis to elaborate a model which assesses, from one side, the warehouse building facilities in accordance with the sustainable development needs and from the other side the way of warehouse equipment and management [9]. Such a model can be applied to compare the sustainability level between different warehouses as well as can be implemented as a tool of self-assessment of the unit aiming the improvement of various aspects of sustainable development.

3 An approach for identification and application of criteria for the purpose of sustainable warehousing assessment model

The final aim of the conducted research is to propose a model, that allows to measure the level of sustainable warehousing. To create such a model it is necessary to indicate the dimensionality of the problem and for this purpose identify the set of criteria which will be the basis for the assessment. An attempt to construct a model has been taken and is presented in [9].

An applied method of model creation is COMET (the Characteristic Objects Method). The COMET method is a multi-criteria decision-making method (MCDM), which is an alternative approach for AHP, TOPSIS, ELECTRE or PROMETHEE methods and allows to eliminate the phenomenon of rank reversal [10]. In accordance with the COMET method 5 steps are required to be carried out [11]:

- defining the space of the problem by determining the dimensionality of the problem (set of criteria) and a set of fuzzy numbers for each of the criteria,
- generating characteristic objects on the basis of Cartesian product,
- determining the Matrix of Expert Judgment to rank and evaluation of the characteristic objects,
- conversion of each characteristic object and preference value into a fuzzy rule.
- inferencing in a fuzzy model and final ranking.

Applying the COMET method lets to create the series of individual models as a result of partial aggregation and finally the group model, which allows to inference on the given problem.

The first step of the method requires to determine the set of criteria and appropriate set of fuzzy numbers for each criteria. The set of fuzzy numbers is presented in natural language. Each fuzzy number is determined by a membership function. The membership function $\mu_A(x)$ describes the membership of the elements x of the base set X in the fuzzy set A.

The functions assigns to elements of X the values from the interval [0,1]. In this case the membership function is a curve that defines how each point in the input space is mapped to a membership value (or degree of membership) between 0 and 1 [12]. It cause that fuzzy sets allow elements to belong to a given set with a certain degree. It is the role of expert to define the fuzzy numbers and appropriate membership functions. In case of sustainable warehousing assessment model this step is crucial for further analysis, because it creates the background for generating the characteristic objects and their ranking and evaluation. The experience and knowledge of experts engage in model creation, and in particular in description of the

As show [12-14], there are many types of membership functions e.g. triangular, trapezoidal, Gaussian, generalized bell, π- shaped and s-shaped membership function. However, the description of criteria applied to create the sustainable warehousing assessment model is based on the trapezoidal and triangular functions. In general, the use of a more complex function does not give increased precision, so triangular and trapezoidal functions are commonly used and quite well adapt to such a problem.

A triangular membership function $\mu_A(x)$ is given as follows [14]:

$$\begin{cases} \frac{x - a_1}{a_2 - a_1}, & for\ a_1 < x < a_2 \\ \frac{a_3 - x}{a_3 - a_2}, & for\ a_2 \leq x < a_3 \\ 0, & otherwise \end{cases} \quad (1)$$

where:

$A = (a_1, a_2, a_3)$

$x, a_1, a_2, a_3 \in R$

$a_1 < a_2 < a_3$

A trapezoidal membership function $\mu_A(x)$ is given as follows [14]:

$$\begin{cases} \frac{x - a_1}{a_2 - a_1}, & for\ a_1 < x < a_2 \\ 1 & for\ a_2 \leq x \leq a_3 \\ \frac{a_4 - x}{a_3 - a_2}, & for\ a_3 < x < a_4 \\ 0, & otherwise \end{cases} \quad (2)$$

where:

$A = (a_1, a_2, a_3, a_4)$

$x, a_1, a_2, a_3, a_4 \in R,$

$a_1 < a_2 < a_3 < a_4$

Application the fuzzy sets and identification the criteria using the linguistic data lets to define the space of considerations of each criteria and simultaneously reduce the dimensionality of further analysis. However, this step require the precisely description of applied measure for each criteria and a given scale as well as such assignment the linguistic values to this scale, that it will properly represent the range of applicable linguistic values.

The other issue is to guarantee the appropriate spectrum of criteria, which will allow inference about the level of sustainable warehousing. From the sustainable development point of view this criteria should include environmental, social and economic dimensions of sustainability. Considering it in case of warehousing, it is required to choose the element regarding to the building design as it is posted in standards as well as technical issues and social policy aspects, which influence on the way the warehouse processes are organized and the resources applied and consumed.

4 Results

Elaborating the model assessing the level of sustainable warehousing was made with the following steps:

- analysis the sustainability in the context of warehouse organization and realization of warehouse processes,
- selecting the set of factors (criteria) influencing the sustainable warehousing based on literature review and expert analysis and considering their impact on economic, environmental and social aspects of sustainable development,
- indicating the measurement scale for each criteria,
- indicating the linguistic scale for each criteria,
- applying the criteria for the purpose of sustainable warehousing assessment model in accordance with COMET method.

The literature review and expert analysis let to identify 22 criteria, which should be taken into account for the purpose of model creation [9]. The spectrum of criteria is presented in Table 1.

Table 1. The range of criteria of sustainable warehouse management assessment model.

Name of the criteria	Meaning of the criteria in the model	Field of warehousing assessment
the use of energy-saving and ecological lighting	the level of certified energy-saving lighting application (eg. LED),	Building design
the use of automatic lighting control devices	the level light sensors application	Building design
the roof surface covered by skylights	the level of roof skylights installation	Building design
light transmittance in skylights	the level transparency glazing (in accordance with the requirements of stored products)	Building design
the use of renewable energy	the level of application the energy from renewable sources comparing to energy from traditional sources	Building design
the use of alternative water sources	the level of rainwater utilization in water consumption	Building design
the use of mechanisms reducing the water flow	the level of sewage points equipped with mechanisms reducing water flow	Building design
the level of building insulation	the thickness of installed insulation	Building design
the use of automatic temperature control devices	the percentage of warehouse area where the possibility of automatic temperature control is given	Building design
the use of air purification methods and tools	the level of purification systems applied in the warehouse	Building design
the use of environmental friendly materials in building structure	the level of the use of pro-ecological materials in the construction of the building	Building design
environmental friendly plot area	the level of biologically active plot area with a diverse flora with low water demand	Building design
level of automation	the level of technological advancement of devices used in the warehouse	Warehouse equipment

IT level	the level of functional complexity of IT systems used in the warehouse	Warehouse equipment
effective inventory control	the level of application the inventory management methods in planning, inbound, storing, picking, and outbound	Warehouse equipment
distance reduction during warehouse operations	the level of application the methods used to orders consolidation, picking paths optimization, picking zones creation	Warehouse equipment
employees' training regarding to recycling and sustainable development	the percentage of employees taking part in training in the field of recycling and principles of sustainable development	Human policy
employees' training regarding to material handling equipment (MHE)	the percentage of employees taking part in training in the field of MHE	Human policy
the use of low-emission moving around the warehouse	the percentage of drivers certified in low-emission driving	Human policy
recycling of packaging and returns	the level of the utilization of returns and recycled materials	Human policy
organization of the employee-friendly zones	number of employees-friendly zones/rooms	Human policy
applying health and safety policy	the level of the quality of the health and safety protection policy	Human policy

The criteria refer to both the technologies and materials used for designing and building the warehouse as well as applied equipment (advancements and application), procedures and standards to promote environmental friendly behaviors and better organization of processes and work conditions.

In accordance with characteristic object method the scale and set of fuzzy numbers was defined. On the figure 1 and figure 2 are presented two examples of the fuzzyfication of two criteria: 1. employees' training regarding to recycling and sustainable development, 2. the level of building insulation. The presented examples concern different issues of sustainability. The first one lets to assess the element of building design and construction and the second one refers to increasing the awareness of people working in warehouse and can be perceived as a part of human policy applied in the organization.

A scale was indicated for each criteria and expressed in natural language using a set of fuzzy numbers. There were three fuzzy numbers in each presented criteria. The membership functions (trapezoidal and triangular) were given in accordance with expert opinion based on his/her knowledge and experience.

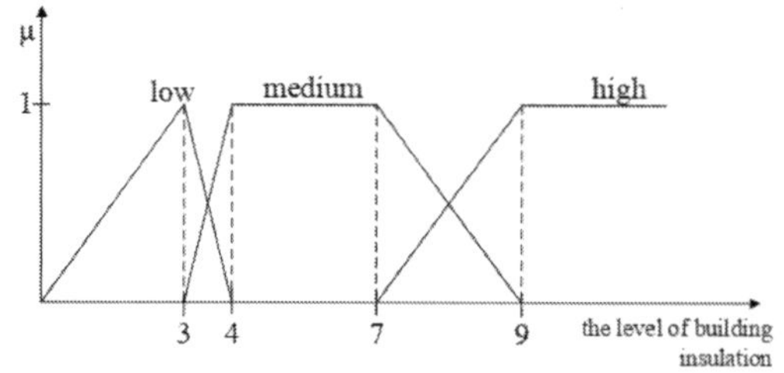

Figure 1. The criteria of level of the building isolation defined in accordance with the fuzzy sets methodology (source: own elaboration)

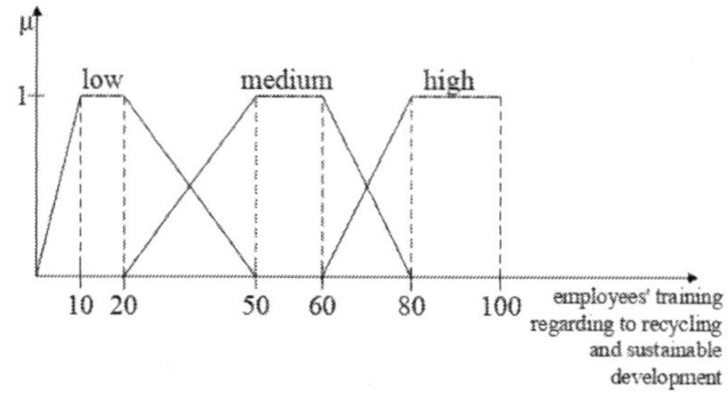

Figure 2. The criteria of employees' training regarding to recycling and sustainable development defined in accordance with the fuzzy sets (source: own elaboration).

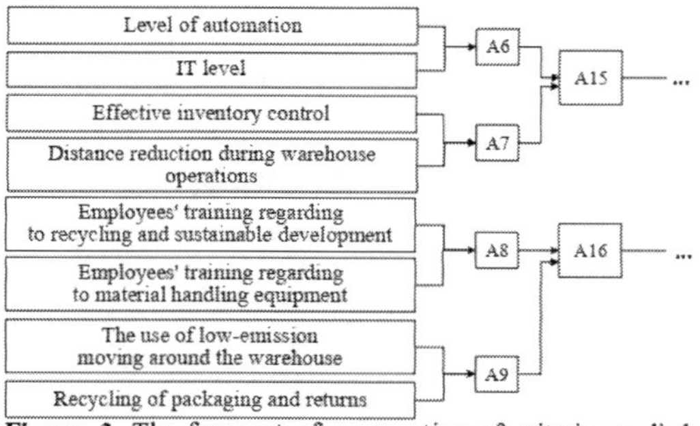

Figure 3. The fragment of aggregation of criteria applied to create the sustainable warehousing assessment model (source: own elaboration)

However the membership function, which determine the fuzzy set is subjective, it makes the foundation for further research. It is possible to verify and change the set of fuzzy numbers used for each criteria as well as the distribution of the membership functions. The changes in this step might be the source of improvement of the sustainable warehousing assessment model.

The method requires, to give the final grade of sustainable warehousing, the creation of successive aggregates built on the basis of selected criteria. The role of the expert is to judge the created object by determining the Matrix of Expert Judgment. In sustainable warehousing assessment model there were established 19 objects. The fragment of prepared aggregation is presented on figure 3.

The object A8, named environmental awareness, was built using the criteria employees' training regarding to recycling and sustainable development (presented on figure 2) and employees' training regarding to material handling equipment. The final ranking allows to determine the level of sustainable warehousing using the 5-elemnts scale expressed in natural language: very low level of sustainability, low level of sustainability, medium level of sustainability, high level of sustainability, very high level of sustainability.

4 Conclusions and Future Research

The growing importance of sustainable development issues causes that in the literature there are many studies which has on aim offering the solutions helping to achieve balance between environmental, social and economic dimensions.

The article presents the original approach to apply the COMET method for the purpose of sustainable warehousing assessment model elaboration. It is indicated, how to utilize the fuzzy set theory to describe the criteria, that are the basis of the assessment and the Matrix of Expert Judgement to reduce the multifurthy of conclusions. The achieved results show that COMET method lets to provide the final model, however, because is based on the expert description of the space of the problem, it is necessary to verify the adopted assumptions of criteria description. The proposed model can be an important step towards finding the measure tool of sustainable warehousing, and the selected criteria may be utilized as indicators showing the directions of improvements implementation to achieve the higher grade of sustainability. The next step of the research is focused to check the complexity of the criteria and the established scales.

Acknowledgments

The article is published with the support obtained under the project 'Improving the quality of scientific research in economic sciences with particular emphasis on the services sector'. The project is financed within the framework of the program of the Minister of Science and Higher Education under the name "Regional Excellence Initiative" in the years 2019 - 2022; project number 001/RID/2018/19; the amount of financing PLN 10,684,000.00.

References

1. Żuchowski W 2015 Division of environmentally sustainable solutions in warehouse management and example methods of their evaluation *LogForum* **11 (2)**, 171-182
2. Płaczek E 2015 Sustainable development in business models of logistics providers *The Wroclaw School of Banking* Research Journal **15 (1)** 87-98
3. Al-Zaidi W A H, Illés Cs B 2017 Dimensions of green warehouses – a literature review, *Selected Collection from Synergy 2017 International Conference* http://www.gek.szie.hu/english/sites/default/files/MEL_2017_16Synergy.pdf
4. Amjed T W, Harrison N J, A model for sustainable warehousing: from theory to best practices. https://pdfs.semanticscholar.org/570e/5476946437853206a3522fd2cad44bb0380f.pdf
5. Polskie Stowarzyszenie Budownictwa Ekologicznego 2016 Certyfikacja zielonych budynków w liczbach https://plgbc.org.pl/wp-content/uploads/2016/05/Certyfikacja-zielonych-budynkow-w-liczbach-2016.pdf
6. Hamedani A Z, Huber F A 2012 comparative study of DGNB, LEED and BREEAM certificate systems in urban sustainability *The Sustainable City VII*, **1**, 121-132
7. Mokrzecka, M 2015 Międzynarodowe systemy certyfikacji LEED, BREEAM I DGNB Wstępna analiza porównawcza poparta studium przypadku *Journal of Civil Engineering, Environment and Architecture*, XXXII **62 (2/15)** 311–322
8. Tan K S, Ahmed M D, Sundaram D 2009 Sustainable Warehouse Management http://ceur-ws.org/Vol-458/paper5.pdf
9. Malinowska M, Rzeczycki A, Sowa M 2018 Roadmap to sustainable warehouse *InfoGlob 2018 SHS Web of Conferences* **57 01028** 1-9
10. Wątróbski J, Sałabun W, 2016 The Characteristic Objects Method: A New Intelligent Decision Support Tool for Sustainable Manufacturing *Sustainable Design and Manufacturing 2016, Smart Innovation, Systems and Technologies* (Switzerland: Springer Int. Pub.) 349-359
11. Sałabun W 2014 The Characteristic Objects Method: A New Distance-based Approach to Multicriteria Decision-making Problems *Journal of Multi-Criteria Decision Analysis* **22 (1-2)** 37-50
12. Ali O A M, Ali A Y, Sumait B S 2015 Comparison between the Effects of Different Types of Membership Functions on Fuzzy Logic Controller Performance, *International Journal of Emerging Engineering Research and Technology* **3 (3)** 76-83
13. Shinde S V, Kulkarni U V 2014 Modified Fuzzy Hyperline-Segment Neural Network For Classification With Mixed Attribues *Fifth International Conference on Computing, Communications and Networking Technologies (ICCCNT) IEEE* 1-7
14. Kabir S, Papadopoulos Y 2018 A review of applications of fuzzy sets to safety and reliability engineering *International Journal of Approximate Reasoning* **100** 29-55

Game theory in creating supply chain logistics strategy - the possibility of applying a holistic approach

A Rzeczycki[1,a]

[1]*Logistics Department, University of Szczecin, Cukrowa 8 Street, 71-004 Szczecin, Poland*

Abstract. The game theory, due to its potential in creating behavioral models, can become a tool used in the area of formulating or redesigning logistic strategies of supply chains (networks). The existing research in this field focuses on the analysis of individual decisions in one specific area. The article discusses the holistic approach to this problem by identifying existing limitations and presenting a framework concept for the application of game theory in the design of supply chain logistics strategies. The conclusions were drawn on the basis of theoretical literature studies, supply chain surveys (on different level chain participants) and laboratory experiment. In results found that the idea of a common interpretation of the strategy and the holistic application of game theory in the search for optimal solutions to decision-making problems requires changing the approach to designing / redesigning the supply chain logistics strategy.

1 The system approach in the analysis of supply chain logistics strategies

The supply chain can be considered by process or systemic approach. The process paradigm operates with such concepts as: process, action, event, process entry and exit, contractor, sequence. The system paradigm focuses on the elements and properties of systems as a whole, together with the structures of connections and occurring couplings indicating that "the whole is more than the sum of its parts, so there is a need to examine individual elements not in isolation from each other, but including the relation taking place between them with specific properties" [1]. As J. Okulewicz points out [2], the process is not a being, but only an observable result of the actions undertaken. According to its nature, it is a dynamic product - it arises and then disappears. It is non-self-made and depends on how it is created by man. From the theoretical side, it is therefore necessary to examine what the process is the result of, and not only elementary activities. The components of the logistics system, which is also the supply chain, at this level of analysis arise not through establishing relationships, but in order to enable satisfying the client's needs. The relations of the supplier or the recipient of goods, forming the supply chain, are created only for the needs of the processes being carried out - they do not constitute the specifics of the logistics system, but only the manner of its functioning. Considering logistic strategy of supply chains in systemic approach, we are dealing with goal-oriented activities (strategy), simultaneous modification of a group of variables (multithreading) and focusing on interactions between elements (decision system) while studying their effects (effects of decisions)

2 Game theory in the analysis of supply chain logistics decisions

Game theory in its assumptions deals with the search for solutions in conflict situations, so the game is a simplified model of the conflict situation, but also possible cooperation. Thus, it apparently inclines to the process approach (concerns relations between players), but taking into account the construction assumptions of the games (rules), which concern, among others: possessed information, communication capabilities, potential results (payoffs) and the usefulness of these results, thoughts are heading towards a systemic approach. It is difficult to analyse the strategy of the entire supply chain. However, the subject of the analysis may remain separated supply chain subsystems concentrated around selected strategic decisions that constitute an elementary part of the strategy's construction. The considerations in the literature in this field include, among others, works on:

- conflicts in the supply chains using non-cooperative and cooperative games in static and dynamic settings, with particular attention to techniques demonstrating the existence and uniqueness of equilibria in non-cooperative games [3],
- conflict and coordination modeling in multi-channel distribution systems belonging to producers and dealers [3],

[a] Corresponding author: andrzej.rzeczycki@usz.edu.pl

 https://doi.org/10.1051/matecconf/201929602006

- bilateral negotiations in supply chains in games with complete and incomplete information, using the perfect equilibrium [3],
- analysis of the supply chain competitiveness; along with the verification of the effectiveness of games measured by the size of market share [4],
- the seller-buyer relationship analyzed using non-cooperating and cooperating games, with Stackelberg's strategy as a concept for a non-cooperative solution and effective Pareto solutions for a cooperative game [5],
- distribution of supply chain risk based on cooperative games, indicating the different ranges of applications of Shapley values, risk distributions in different operating conditions, and the proposal of two methods for distributing the decreasing risk of cooperation [6],
- benefits resulting from sharing information on demand in conditions of cooperation and competition in the supply chain of one supplier and one retailer, considered in various decision-making modes [7],
- distribution of profits in the supply chain in relation to the advertising activities undertaken, using the equilibria of Nash and Stackelberg and the search for an effective point in the sense of Pareto [8],
- profit sharing in a two-cycle supply chain consisting of one risk-neutral supplier and two risk-prone retailers, when players do not cooperate and cooperate with each other, and retailers face the problem of stochastic price-sensitive demand [9],
- application of Stackelberg's theory for modeling and analyzing a supply chain system consisting of a single supplier and many retailers using the inventory management method by the supplier, by presenting a general non-linear mixed model that explains revenue sharing, price agreement for wholesale prices and a centralized structure [10],
- optimization of logistics costs by modeling interaction between players in a single-channel, two-way supply chain, paying particular attention to sharing information and forecasting demand [11],
- coordination of a two-cycle supply chain consisting of a producer and seller with an element of joint advertising and price decisions, in which the seller performs local advertising, and the producer pays for some or all of the cost of advertising to the retailer, using Stackelberg's balance with asymmetrical distribution of power [12],
- planning production processes to minimize the function of production and transport costs by seeking a balance point using clean and mixed strategies [13],
- analysis of the behavior of the participants in the supply chain in order to examine the possibility of coordination through incentives, due to the conflict and non-participation of players in the network design process, using the so-called social utility [14],
- application of game theory in negotiations, in various tender conditions, use cases and logistic processes [15].

Determining stimulants and destimulants of the application of game theory in rationalization of supply chains, the subject should be analyzed from the theoretical and empirical side. When assessing the applicability from the theoretical side, it is necessary to point to a large number of methods, techniques and tools affecting the extent of potential applications. It is both a disadvantage and an advantage. On the one hand, it makes it possible to consider practically every problem in different conditions, on the other hand it requires a very good knowledge of the subject, in order to select the right analysis instrumentation. Evidence of how many approaches and methods was developed by developing the theory of games is the analysis carried out at the beginning of the 21st century as part of the GAMUT project, during which 122 different sets of games were identified for which 71 generative procedures were found, and finally 35 basic classes of games were identified in order to creating an IT tool that allows testing any algorithm that requires a game in its normal form as an entry [16].

Knowledge of the games themselves does not exhaust the topic of the complexity of the application of game theory. The game should be selected based on the rules, the planned set of actions and the approach to the solution. In practice, this means that, for example, when considering auctions, presenting the game and choosing how to solve it will depend on such parameters as: number of sellers and buyers, type of auction (forward, backward, double, etc.), complexity of offers (simple, related), agents' preferences, organization of auctions (one-off, multiple, parallel, sequential, etc.), possibilities of cooperation, product characteristics (eg divisible, indivisible, structured), availability of information, and many others.

The difficulty in applying game theory in the rationalization of logistics strategies is also the problem of the construction of the payment function. This applies to the transfer to a numerical expression (profit, cost, price, etc.) of issues such as product quality or service, timeliness, flexibility, and even going further, symbolic values such as trust or experience.

These problems effectively destimulate the application of game theory, in particular - in economic practice. The solution is to create information management systems based on these concepts. However, while it is possible to prepare algorithms and even a fairly simple database of rules, the payout structure will be an obstacle, which due to the number of possible factors must be assessed by the decision maker (for example a weighted averages system can be used).

However, before one can talk about the possibility of using game theory in IT systems of logistic management, the theoretical basis of its functioning should be properly prepared. Game theory methods have many opponents due to their limitations and logical simplifications. A huge application problem of game theory are solutions derived from mathematical rules, which for obvious reasons (eg economic) should not be chosen. For example, a coordination game in which two players can choose a "red" or "blue" strategy earns 10 units if both players choose the "red" option and 5 units if both choose "blue". In the options of different strategies, their payout is less than 5 units. In this case, the obvious solution (with full information) is the "red" strategy, but the game theory points to two Nash equilibria. The option to choose the "blue" strategy is also universally accepted for the wet-for-wet strategy. Although this strategy is better than other solutions in the iterated dilemma of the prisoner, its evolutionary stability is undermined. It turns out that it will get the highest payout, if players do not discount the future too much and enough of them use the "tit for tat" strategy [17].

There are many more such problems, but in the context of the considerations on the creation of logistics strategies and relations in supply chains, the nature of player behavior and the rationality of the choices made by them should be considered the most important. Admittedly, game theory offers adaptation solutions for limited player rationality considering behavioral factors, but at the same time allows for assumptions such as symmetry in which agents can't distinguish between themselves if they are assigned to a symmetrical negotiation problem or the same perception of the game by all players . Of course, the idea of the model is to simplify reality, but it is difficult not to incline to the conclusion that this is too important factor to exclude it from analysis, especially as experience indicates that the application of game theory in practice often does not lead to anticipated events. Therefore, key questions arise: how strongly are the behavioral factors influencing strategic decisions in the supply chains, whether they depend on other modeled variables, or the probability of deviation is large enough to be considered and whether the assumptions regarding the rationality of human behavior are they bigger than it is possible to obtain?

Although the utilitarianism of game theory raises many doubts, it must be admitted that it also introduces taxonomy in this field, and despite the difficulties in achieving a clear solution, it allows building general behavior patterns supporting strategic logistics decisions. It also allows for the consideration of potential solutions that are ignored by other theories supporting decisions making..

3 The framework concept of constructing supply chain logistics strategies using game theory

The supply chain strategy depends on the bargaining power of the entire supply chain, which depends on the bargaining position of its participants. Further, the behavioral theory of games and managerial theory of the enterprise add to the requirements that should be taken into account in formulating strategies, managerial goals that are conditioned by motives, personality traits and cognitive abilities of managers. To verify these factors, surveys and a laboratory experiment were carried out. In results found that the idea of a common interpretation of the strategy and the holistic application of game theory in the search for optimal solutions to decision-making problems requires changing the approach to designing / redesigning the supply chain logistics strategy.

In the literature of the subject man can find many schools of strategies designing. Consideration of decision elements leaning towards use of the school of power and culture, and a fragmentary approach to the problems under consideration, towards the configuration school. According to the school of power and culture, strategy design is treated as a process of negotiation through the clash of different stakeholder groups, as well as a process of social impacts based on values and stakeholder standards. Strategy design according to the configuration school is treated as a transformation process. This school assumes the stability of the strategy during a given state of the unit, interrupted by occasional transitions to the new phase [18].

The combination of these approaches will allow the creation of a supply chain logistics strategy in a modularized approach. In other words, the starting point is not so much the mission and vision of the chain, but the dynamically interpreted bundle of goals and strategic rules. The objectives of variables, as they result from the bargaining positions of individual links in the chain and the strategic opportunities resulting from their competence. The strategic rule (this phrase refers to the rules of the game) should be understood here as the superior method of hierarchizing goals. For example, the agile strategy rule will prioritize solutions that accelerate and make functioning more flexible, but do not eliminate solutions that reduce costs. Here will work the trade-off principle. Such an approach is closer to the tendencies identified in practice for the emergence of mixed strategies and the concepts of managing managerial goals in various areas of the chain and its participants.

The formulation of a logistics strategy for the supply chain is therefore carried out in a three-tier manner. Two tracks deal with internal processes in each of the participants in the supply chain, while the third involves a confrontation of the internal assumptions of all links. According to this concept, the company assesses its strengths and weaknesses by identifying key competences that, under the influence of stakeholders' goals (especially owners expecting value growth and profits and customers who set specific requirements), are transformed into strategic rules and frameworks of business and functional strategies. At the same time, the decision maker, i.e. the manager, analyses the environment (internal supply chain and out of the chain), identifying opportunities and threats related to the uncertainty of functioning in its decision area. Acting in accordance with its goals, as well as knowledge and experience, it imaging the chain, and thus evaluating its strategic (negotiating) position towards other links. On this basis, taking into account the developed strategy framework and relationships in the chain (cooperation, competition, control), each manager in his functional area, looks for opportunities and makes decisions. These decisions (ultimately using the assumptions of game theory) cause the reconstruction / redesign of the previous assumptions of the business and functional strategy. Such a process, consciously or less, can take place at every link in the chain, influencing the relationships being built and the emergence of strategic opportunities, as well as the shaping of the logistics strategy of supply chains..

4 Conclusion

The use of game theory in the design / redesign of supply chain logistics strategies allows the economic rationalization of these chains by estimating and indicating decision paths that ensure an increase in profits, a decrease in costs, or an improvement in resource management. Achieving such results is possible thanks to a thorough analytical procedure, taking into account the player's behavior, which allows more accurate estimation of the probability of events, and even (in favorable conditions) creating the behavior of game participants.

Rationalization can be achieved both at the internal level of the organization (logistic aspects of business strategy, functional strategy) and external (supply chain strategy). There are, however, some differences in the approach to games. At the internal level, games are less complicated because they usually involve fewer players, and activities are subordinated to the overall strategic concept. The external level is regulated by relationships of various kinds, and the heterogeneous goals of chain participants and the way the managers perceive the chain increase the number of likely interactions. Based on the conducted research, there were also discrepancies in the priorities of action at individual organizational levels independent of the declared strategies.

The conducted research also allowed for drawing a number of additional applications of a utilitarian nature, which include following:

- game theory is a multi-faceted field, in which the research has broad application potential for analysis of conflicts occurring in the supply chain, particularly in the area of trade negotiations, contracts and the distribution of payment,
- the applicability of game theory in modeling supply chain logistics strategies is determined by the variety of possible conflict situations (payout rules and functions) and the approaches to solving them developed in game theory,
- occurring strong behavioral aspects introduce information disruptions to the supply chain games,
- the level of applicability of game theory increases for iterative games where the conditions remain the same,
- empirical studies show the existence of a weak correlation between the commonly recognized factors in building a logistics strategy and the strategy implemented in individual links in the supply chain, which indicates the existence of other "forces" affecting the shape of the strategy,
- the identified market problem, as well as the threat of a modular approach to building logistics strategies is adhesiveness, instead of cohesion (coherence) of strategies at various levels of their functioning,
- the manager's perception and competences determine his strategic choices, becoming the foundation of the strategy and a representation of the network environment; therefore, studying the representation of the environment that shapes the rules of decision games is the key for understanding strategic choices,
- the use of a holistic approach to game theory in the analysis of decision problems allows the elimination of insignificant alternatives, which take the form of relatively stable results, with changing game rules.

Acknowledgments

The article is financed within the framework of the program of the Minister of Science and Higher Education under the name "Regional Excellence Initiative" in the years 2019 - 2022; project number 001/RID/2018/19; the amount of financing PLN 10,684,000.00

References

1. Krupski R., Przegląd Organizacji, **4**, 3-7 (2012)
2. Okulewicz J., Prace Naukowe Politechniki Warszawskiej, **64**, 103-110 (2008)
3. Simchi-Levi D., Wu D.S., Zuo-Jun M., Handbook of Quantitative Supply Chain Analysis (Springer, Berlin, 2004)
4. Ming Q., Wen-fang S., Yun-ling Ch., *Proceedings of International Conference on Management Science and Engineering* (Harbin, China, 2007)
5. Esmaeili M., Aryanezhad M.B., Zeephongsekul P., Eur. J. of Operational Research, **195**, 442-448 (2008)
6. Xiaofen J., Wei Ch., Chen P., Rongrong X., *Proceedings of the International Symposium on Intelligent Information Systems and Applications* (Qingdao, China, 2009)
7. Tetsuo I., Zipkin P., J. of Operations Research, **58**, 1350-1363 (2010)
8. Ahmadi J.A., Hoseinpour P., J. of Optimization Theory and Applications **149(1)**, 138–150 (2011)
9. Changwen L., Yongwu Z., Yong L., Jinsen G., J. of Ind. Eng. and Management, **4**, 816-830 (2014)
10. Rasay H., Zaremehrjerdi Y., Fallahnezhad M.S., Int. J. of Supply and Operations Menag., **4**, 439-465 (2015)
11. Slimani I., El Farissi I., Achchab S., J. of Theoretical and Applied Information Technology, **82(3)**, 411-416 (2015)
12. Kazemi A., Saeedmohammadi Z., Int. J. of Supply and Operation Management **2(4)**, 1035-1063 (2016)
13. Oleander M., Skołud B., Krenczyk D., Przedsiębiorczość i zarządzanie, **17(12)**, 295-307 (2016)
14. Panchala G., Jainb V., Cheikhrouhouc N., Gurtner M., J. of Revenue and Pricing Management **16(4)**, 417–436 (2017)
15. Škrinjar J.P., Abramović B., Furdić M., MATEC Web of Conferences, **134(00053)**, 1-8 (2017)
16. Nudelman E., Wortman J., Shoham Y., Leyton-Brown K., *Proceedings of the Third International Joint Conference on Autonomous Agents and Multiagent Systems* (New York, 2004)
17. Bendor J., Świstak P., *Studia socjologiczne,* **3(150)**, 127-171 (1998)
18. Sołtysik M., *Zeszyty Naukowe Uniwersytetu Ekonomicznego w Krakowie,* **910**, 21-35 (2013)

Total cost of ownership in pallet pools management

Mariusz Sowa[1,a]

[1]*Department of Logistics, Faculty of management and Economics of Services, University of Szczecin, Cukrowa 8 Street, 71-004 Szczecin, Poland*

Abstract. Every loading pallet, regardless of the type, in order to be considered safe during the operation phase, must meet certain formal requirements already at the implementation stage. The loading pallet marketed has a certain life span. The life of the pallet depends on many factors. First of all, they are related to the physical and economic aging of the product. Physical aging consists in the fact that the materials from which the products have been made lose their properties, and the economic ones are the result of the emergence of new, better products on the market, better meeting the specific needs of recipients. The operational life of the cargo pallet also depends on the organization of the transport process, branch and means of transport as well as the date of its completion. The concept of calculating the total cost of ownership of a reusable loading pallet in logistic processes, proposed in the article, allows to determine the total costs of its operation. It is based on the idea of Total Cost of Ownership (TCO), which includes total acquisition costs, installation costs, maintenance costs and costs of its disposal over particular time

1 Introduction

A pallet is most often a wooden structure that is used for transporting goods / loads, but also finds use as a rigid base on which larger-sized goods are laid and in smaller packages, thus facilitating logistic operations. The pallets are designed by manufacturers so that they can be easily lifted using forklifts and other logistics equipment. Currently, the pallet market is striving for a greater variety of pallets, both in terms of dimensions, materials from which they are made, and due to the type of their use. Pallets should be considered as an important element of the comprehensive logistics process. Optimally selected and used pallets contribute to both lowering the costs of freight flows and to increasing the quality of logistics processes.

A pallet is a supporting transport and storage equipment, adapted to motorized facilities and for stacking loads. Pallet load units created on the basis of various types of pallets can be used in rotation taking place in numerous actors of chains of supply. To be considered safe in the operating phase each pallet, regardless of its type, has to meet particular formal requirements as early as at the workmanship phase. However, the operating phase of a product can be defined as its total duration on the market, which starts when the product is released on the market until it is withdrawn from it. Therefore, a pallet released on the market has service life. This, above all, is related to economic and physical ageing of the product. Whereas physical ageing consists in materials used to manufacture the product that lose its properties, economic ageing results from the introduction of new, better products which are better targeted to the needs of consumers. Service life of a pallet depends on the way transporting process is organized, mode and means of transport and time when it was produced as well.

Possession of ownership obviously involves purchase costs, but ownership also involves costs associated with the installation, implementation, use, updating and maintenance of the same resources. These costs after purchase can be significant. Therefore, for many types of assets total cost of ownership analysis of TCO, finds a significant difference between the purchase price and the total life-cycle costs. The difference can be very large when the ownership period covers a longer period. As a result, the TCO analysis sends an important message to corporate buyers, capital control groups and asset managers. Total Cost of Ownership (TCO) is an analysis aimed at discovering all life-cycle costs resulting from having certain types of assets. As a result, the English-language literature TCO is often called life-cycle cost analysis.

2 2. The life cycle of the pallet load

The physical aspect of the life cycle of a loading pallet can be understood as a material based life cycle, which is based on the transformation of materials and energy. Is a sequence of activities such as: acquisition and purchase of raw materials, material processing, production (assembly) of pallets, use and distribution of pallets, recovery and utilization of pallets. In the life cycle understood in this way, the loading pallet begins its "life"

[a] Corresponding author: mariuszsowa.ps@poczta.fm

from the moment of obtaining the raw material necessary for its production.

The assessment of the actual lifetime of the pallet is possible due to the analysis of the conditions of its operation in particular branches of transport. However, the logistics life cycle of individual pallets, since its introduction until the withdrawal from the market, in addition to the time factor should be analysed sales volume and the relationship between the quality of the product and the requirements of consumers, as well as issues of efficiency of its use in supply chains. In order to assess the actual duration of the operational phase of the loading pallet in particular branches of transport, it is necessary to analyse the type of transhipment devices used, the means of transport used, atmospheric conditions, duration of displacement and the mode of its operation. The type of material raw materials, their composition, proportions, methods and techniques of production and the intended purpose of the product have an impact on its life cycle [1]. However, shortening the product life time is not always the result of a decrease in its durability or an earlier loss of its technical and utility properties. The introduction of new or modernized products may also contribute to this phenomenon. However, this may lead to an increase in the ecological hazard caused by the operation of products, the emergence of post-consumer and production waste as well as the increased acquisition of primary raw materials and energy. Acquisition of primary raw materials, their processing and production of final products cause negative changes in ecosystems, change living conditions and disturb the balance of the ecosystem. Processes occurring in production systems are the source of waste with various concentrations and harmfulness.

The total cost of ownership highlights the difference between the purchase price and the long-term cost. This analysis has been at the center of interest since the mid-1980s due to expenses related to the purchase of IT hardware and software. Managers then discovered that hardware and software support could cost 5 to 8 times more than the purchase price.

When the differences between total cost of ownership and price were at the forefront, companies began to use these calculations for a wide range of investment decisions, including: buildings, vehicles, production equipment and information technology infrastructure. The concept of the total cost of ownership of a reusable flat load pallet allows to determine the total costs of its use and to indicate activities that increase its operational life. It is based on the idea of Total Cost of Ownership (TCO), which includes the total cost of obtaining, putting into use (installation), use, maintenance and finally getting rid of the product over a specified period of time. The proposed TCO concept contributes to the recognition of relatively all costs associated with the entire lifetime of a reusable loading pallet. It also allows you to better understand and account for future costs that are invisible when you buy a pallet. The cost of buying a pallet is relatively simple to estimate. However, the calculation of the costs of the pallet's design, its assembly, maintenance, modernization and finally withdrawal from use is much more difficult. Helpful in this situation may be to use the model of the total cost of ownership. TCO plays an important role in making informed financial decisions regarding the purchase of a given product or service. In the case of loading pallets, it takes into account not only the purchase price of the pallet, but also the anticipated costs related to its life cycle [2], which include, among others: costs related to the design of the pallet, the purchase cost of the pallet, operating costs of the pallet, maintenance costs pallets. From the factors mentioned above, it follows that the total cost of owning a pallet is not only the cost of buying it, but also the costs associated with its maintenance. The pallet's TCO model is used to assess both current and forecast costs. It includes direct costs as well as indirect costs (hidden) that are revealed during the operation phase. The TCO model can be divided into individual groups of costs related to the life cycle of the pallet, as shown in Figure 1.

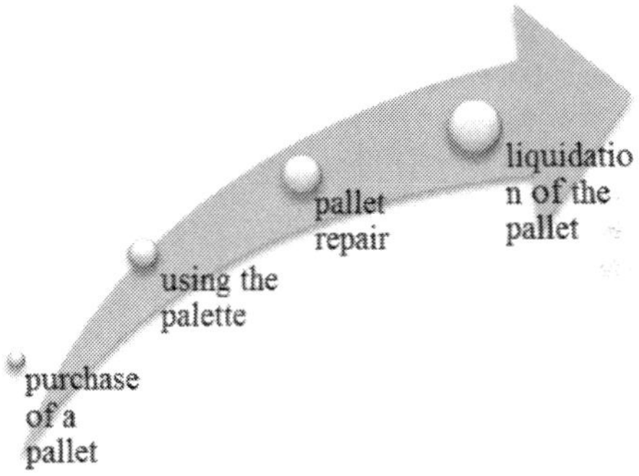

Figure 1. Simplified life cycle of the loading pallet Source: Own study

On the basis of this life cycle loading pallets can be distinguished among others cost group:

- costs related to the purchase of a pallet, which include, among others[3]:
  - direct costs related to production, such as, among others:
    - design costs (costs of market research, costs of prototyping, testing costs),
    - costs of using direct materials (wood), components,
    - costs of purchased materials (nails, brackets),
    - direct labor costs,
    - costs of energy and fuel consumption,
    - costs of processing a foreign raw material, if it is not carried out by a production company. Wooden elements, pallets, dunnage materials made of wood must be subjected to a series of processes, such as: heat treatment, fumigation, certification and marking according to the ISPM15 standard,
    - costs of remuneration, insurance and other benefits of production workers and production supervision (apprenticeship allowances, bonuses, holiday pay),
  - indirect costs associated with the production, such as[4]:
    - faculty costs:
    - fixed - costs remaining unchanged despite fluctuations in production volume, e.g.

depreciation of machines, devices, buildings and structures, non-life insurance, costs of heating, lighting, maintenance, etc.,

- variables - costs depending on the size of production, e.g. drive energy of machines and devices, maintenance and repair costs of machinery and equipment, consumption of materials for traffic, etc.,
- board costs[5]:
- administrative and economic costs, for example: wages of management and administration employees along with overheads,
- business travel costs of management and administration employees,
- office costs, such as office supplies, telecommunications charges, costs of data archiving, accounting services,
- training costs for management and administration staff,
- cost taxes, such as a property tax on the board building
- general production costs [6]:
- maintenance costs for buildings and general premises as well as management,
- costs of the usage license,
- costs related to the use of the pallet, related to the current operation of the pallet, including[7]:
  - acquisition cost,
  - storage cost,
  - transport costs,
  - servicing costs,
  - costs related to obtaining certificates,
  - costs of pallet exchange materials,
  - repair costs of the pallet,
- costs related to the liquidation of the pallet, which include:
  - transport costs,
  - disassembly costs,
  - disposal costs.

3 The total cost of ownership of the loading pallet

Obviously, ownership of property is associated with purchase costs, but owning also has costs associated with installing, implementing, using, updating and maintaining the same resources. These costs can be significant after purchase. Therefore, for many types of assets, the analysis of the total cost of ownership of a TCO finds a significant difference between the purchase price and the total life cycle costs. The difference can be very large if the ownership covers a longer period. As a result, the TCO analysis sends an important message to corporate buyers, capital control groups and asset managers. Total Cost of Ownership (TCO) is an analysis aimed at uncovering all life-cycle costs arising from having specific types of assets. As a result, in English literature on the subject, TCO is often called life-cycle cost analysis. Total Cost of Ownership (TCO) is a term that is intended to determine the structure of costs associated with the operation of assets in a given enterprise. Refers to assets (investments) with long-term economic consequences. TCO is a method, often understood as a model, a concept that allows the decision-maker not only to determine the direct costs that are reflected in accounting documents (invoices), but also to estimate indirect (hidden) costs calculated according to individual division keys used in a given business entity. It can be used to estimate the costs that an investor will have to bear during the operation of a given asset (e.g. car, garage, warehouse, IT system, pallet), but also allows (at the stage of making a decision on purchasing or manufacturing) to determine what will be loss of market value (for example, after a given useful life of the asset. It is a concept of total costs of use, which presents a comprehensive approach to the costs of a given technology, e.g. IT or a given investment (e.g. purchase of pallets) in a given time period. The TCO model also includes a methodology for planning and conducting continuous cost rationalization. TCO contributes to knowing relatively all costs that are associated with the entire lifetime of the pallet. It also allows you to better understand and account for future costs that are invisible when you buy the pallet. The cost of purchasing a pallet as a carrier is relatively simple to estimate. However, calculating the costs of pallet design, assembly, maintenance, modernization and finally decommissioning is much more difficult. The total cost of ownership model may be helpful in this situation. TCO plays an important role in making informed financial decisions regarding the purchase of a given product or service. In the case of pallets, it not only takes into account the purchase price of the pallet, but also the anticipated costs related to its life cycle.

The total cost of ownership of the loading pallet consists of the sum of: purchase cost, maintenance cost, operating cost and withdrawal cost. Therefore, the TCO formula of a reusable flat pallet takes into account, in addition to the purchase price of a pallet, service or solution, also the expected costs related to the life of the pallet, which include servicing, repairs, storage, etc. This concept is presented in the figure 2.

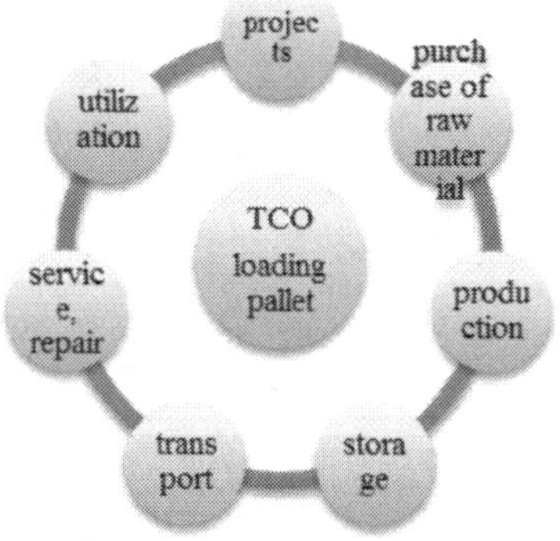

Figure 2. General assumptions of the TCO model of the loading pallet
Source: Own study.

The factors presented in the figure show that the total cost of owning a pallet is not only the cost of purchasing it, but also the costs associated with maintaining it. The TCO model of the pallet is used to evaluate both current and projected costs. It includes direct costs as well as indirect (hidden) costs that are revealed during the exploitation phase.

TCO model shows all the costs that are incurred throughout the life cycle of the pallet, since her shopping until getting rid of. If you add the costs of storage, repair costs or the cost of disposal to the basic price of the pallet, it may turn out that a much more advantageous solution for the company will be to use the palette pool system (open pool palettes) than definitely to buy it [8]. In taking this decision will serve analysis of the total cost of ownership, thanks to which the company is able to provide for additional costs associated with the purchase of the pallet.

At the stage of making decisions regarding the purchase of pallets by the company, the criterion that is taken into account is the most common and only the price. Entrepreneurs rarely include other expenses related to owning a pallet pool, such as e.g. repair costs. However, to have a full picture of it, it becomes justified to use the TCO model, because it indicates all the costs that are incurred throughout the entire life cycle of the pallet, from the moment of its purchase until disposal. If we add storage costs, repair costs or disposal costs to the basic price of the pallet, it may turn out that using a pallet rental system (open pallet system) will be a much more beneficial solution than definitely buying it. This decision will certainly be based on an analysis of the total cost of ownership, thanks to which the company is able to predict the additional costs associated with purchasing a pallet.

4 Conclusion

Total Cost of Ownership (TCO) is a term used to determine the structure of costs related to the operation of assets in a given enterprise. It applies to assets (investments) on the long-term economic consequences. TCO is often understood as a model, a concept that allows the decision maker not only to determine direct costs, which are reflected in accounting documents (invoices), but also to estimate indirect (hidden) costs calculated according to individual division keys used in a given economic unit. It includes the total cost of obtaining, putting into use (installation), use, maintenance and finally getting rid of the product over a specified period of time. The proposed TCO concept of the loading pallet contributes to the recognition of all the costs associated with the entire life of the pallet. It also allows you to better understand and account for future costs that are invisible when you buy a pallet. The pallet's TCO model is used to assess both current and forecast costs. It includes direct costs as well as indirect costs (hidden) that are revealed during the operation phase. This concept will allow enterprises to obtain a useful instrument for determining the total operating costs of a pallet that is widely used in the course of trade. It will also allow to indicate the area of activities increasing its operational life, which will result in their better use, cost savings, but also reduce the negative impact of their logistics activities on the natural environment.

Acknowledgments

The article is financed within the framework of the program of the Minister of Science and Higher Education under the name "Regional Excellence Initiative" in the years 2019 - 2022; project number 001/RID/2018/19; the amount of financing PLN 10,684,000.00

References

1. Rosa G Perenc J Ostrowska I 2016 *Marketing przyszłości Od ujęcia tradycyjnego do nowoczesnego Wydawnictwo* C. H. Beck Warszawa pp.171-173
2. Selech J Kurczewski P 2010 *Metoda szacowania kosztu cyklu życia (LCC) i jej zastosowanie w dziedzinie budowy i eksploatacji obiektów technicznych* Inżynieria i aparatura chemiczna nr 5 pp.79-83
3. Sobańska I 2009 *Podstawy rachunku kosztów i wyników* W *Rachunek kosztów Podejście operacyjne i strategiczne* Wydawnictwo C.H. Beck Warszawa pp.80-87
4. Sobańska I 2006 *Rachunek kosztów i rachunkowość zarządcza Najnowsze tendencje, procedury i ich zastosowanie w przedsiębiorstwach* Wydawnictwo C.H.Beck Warszawa pp.119-127
5. Szczypa P 2017 *Zasady rachunkowości Teoria przykłady zadania* CeDeWu, Warszawa pp.152-153
6. Okoń J 2014 *Zastosowanie rachunku kosztów działań w zarządzaniu portfelem produktów* W Studia ekonomiczne Kierunki harmonizacji polskiej rachunkowości z międzynarodowymi standardami sprawozdawczości finansowej Seria Zeszyty Naukowe Wydziałowe Uniwersytetu Ekonomicznego w Katowicach Katowice pp.45-49
7. Jarugowa A Malc W Sawicki K 2012 *Rachunek kosztów* PWE Warszawa 1983 pp.44, cyt. za Szczypa P *Rachunkowość zarządcza. Klucz do sukcesu* CEDEWU Warszawa pp.20-21
8. Jedliński M 2016 *Koncepcja dynamicznej efektywności palety w analizie społecznych kosztów i korzyści w fazie użytkowania palety drewnianej wielokrotnego użytku* [III Konferencja Paletowa PKN EPAL, Józefów

Quality Engineering with Taguchi Loss Function Method and Improvement of Work Method in Anode Changing

Ikhsan Siregar[1,a]

[1] *Departement of Industrial Enginnering, Faculty of Engineering, Universitas Sumatera Utara*

Abstract. One of the companies engaged in aluminum smelting production has problems, namely the occurrence of high variations in the removal of new anodes to the old anodes which are influenced by the operator's working method because of unnecessary movements and movements beyond the Anode Changing mechanism set by the company has become a habit of the operator itself, so the time needed to replace the anode is getting longer. This problem causes the company to suffer losses due to the production process that runs not in accordance with established procedures. So, the company needs to calculate the losses incurred due to variations in anode replacement and see how the operator's working method when changing the anode that causes these variations and also has an influence on the anode replacement cycle time. The method used for this problem is Taguchi Loss Function which is used to calculate the losses suffered by the company and improve work methods using the Modular Arrangement of Predermined Time Standards (MODAPTS) method to calculate the processing time. The results obtained using this method are losses experienced by the company in Block 1 amounting to Rp 19,733,263, while the total losses in Block 2 amount to Rp 35,919,435. While for the operator's working method, the operator's work movements are analyzed which are not in accordance with the economic principles of the movement. Comparison of the standard time of actual and proposed work methods is 774.847 seconds and 648.2 seconds with a time difference of 126,647 seconds faster and there are no movements that should not be done so that the operator can use the time to adjust the anode height as well as possible according to the standards determined by the company.

1 Introduction

The design of the production process is one part of quality engineering activities. Production processes that pay attention to quality will produce good quality processes so that they can avoid losses in production costs, processing time and others. Quality control is an activity to ascertain whether the policy in terms of quality (standard) can be reflected in the final result, or in other words an effort to maintain the quality or quality of the goods produced to conform to the product specifications determined based on the leadership policy [1,2]. The smooth production process is strongly influenced by the production system that has been prepared before the company carries out the production process. In addition, for the sake of the smooth production process, it is also necessary to control the production process which will control all important components in a company [3]. The production process to produce a kind of output is difficult to avoid the occurrence of variations in the process of defining variation as a tendency in the production or operational system so that the difference in quality at the output (goods and services produced) [4,5]. To solve these problems, this study will use the Taguchi Loss Function method. Taguchi has the view that quality is related to costs and losses in the monetary unit. Losses suffered include the production process and losses suffered by consumers [6,7]. Quality is anything that can satisfy customer desires or according to customer requirements and requirements [8]. The purpose of the Taguchi loss function (loss function) is to evaluate quality losses quantitatively due to variations [9]. In analyzing and evaluating work methods to obtain a more efficient working method, it is necessary to consider the economic principles of the movement. The economic principle of this movement can be used to analyze local work movements that occur in a work process and also for work activities that take place comprehensively from one process to another [10].

[a] Corresponding author: ikhsan.siregar@usu.ac.id

This method is widely used in research, one of which is research conducted by Antony Jiju (2001) regarding Simultaneous Optimization of Multiple Quality Characteristics in Manufacturing Processes Using Taguchi's Quality Loss Function. The methodology uses the Taguchi's quality loss function for interaction effects and also for determining the optimal condition of the process. In order to demonstrate the proposed methodology, a simple case carried out for optimizing three quality characteristics, namely solder paste mass, solder paste height, and glue torque, for a double-sided surface mounting technology, electronic assembly operation. Six control factors and interaction effects were studied using an L8 OA experiment advocated by Taguchi [11]. In addition, Irena Šabarić, Snjenaza Brnada, Stana Kovacevic (2013) conducted research entitled Application of the MODAPTS Method with Innovative Solutions in the Warping Process, located in Zagreb-Croatia. This study aims to eliminate ineffective movements in the process of compiling yarn rolls in textile companies by designing innovative work aids. The results of the study state that from the motion study calculation using MODAPTS can save time by 75.87% mods after repairs due to the simplification of work movements using innovative trolleys [12].

Taguchi Methods involve reducing variations of the process through the robust design of the experiment. The main objective of this method is to produce high quality products at very low costs. Taguchi developed a method for designing experiments to be able to investigate the effect of different parameters on the mean and the variance of the performance characteristics of the process that determines how well the process is functioning. The experimental design introduced by Taguchi involves orthogonal arrays to organize parameters that influence the process and levels that need to be varied. Taguchi Methods do not test all possible combinations but only test a few combinations. This test will produce a collection of important data that can determine what factors most influence product quality with minimum experimentation to save time and money.

2 Methodology

The research was conducted at one of the factories engaged in aluminum smelting production. The object of the researchers observed was the changing anode and the operator's working method where the research was conducted in August 2015. The type of research design used was a type of causal research. This is because this study was conducted to investigate causal relationships by observing the consequences that occur and the possible factors (causes) that cause these effects [13]. Problem-solving is done by calculating the Loss Function of the variations that occur, then making improvements to the work method.

Consumer products and services are now increasingly critical of the products offered by producers. In addition to paying attention to the prices offered, more attention to the quality of goods or services offered by producers. Therefore, producers must also pay attention to the quality of products or services offered to consumers in order to maintain or expand market share.

There are several types of quality control, including:

1. Inspection
2. Quality control with the Taguchi Method
3. Statistical quality control

The general steps in Taguchi Methods are as follows:

- Determine the purpose of the process or more specifically the target value for measuring the performance of a process.
- Determine the design parameters that influence the process
- Make orthogonal arrays to design parameters that indicate the number and conditions of each experiment
- Connecting experiments indicated on arrays that have been completed to collect data on the effects of performance measurements
- Complete data analysis to determine the effect of various parameters on performance measurement.
- The Taguchi Method is best used when there are intermediate numbers of variables (3 to 50), small interactions between variables, and when only a few variables make a significant contribution.

In Taguchi Method, the word "optimization" implies "determination of BEST levels of control factors". In turn, the BEST levels of control factors are those that maximize the Signal-to-Noise ratios. The Signal-to-Noise ratios are log functions of desired output characteristics. The experiments, that are conducted to determine the BEST levels, are based on "Orthogonal Arrays", are balanced with respect to all control factors and yet are minimum in number. This in turn implies that the resources (materials and time) required for the experiments are also minimum.

Taguchi Method is a process/product optimization method that is based on 8-steps of planning, conducting and evaluating results of matrix experiments to determine the best levels of control factors. The primary goal is to keep the variance in the output very low even in the presence of noise inputs. Thus, the processes/products are made ROBUST against all variations.

2.1. Steps

The data processing stage that is carried out is:

a. Calculating Loss Function from variations that occur using anode and AVV height data (Average Voltage)
b. Improvement of work methods
   1. Map of the left hand and right hand
   2. Modular Arrangement of Predermined Time Standards (MODAPTS)

3 Result and Discussion

3.1. Calculating Loss Function

Taguchi defines quality as a loss in a society starting from the delivery of a product. Losses include expenses, waste, and opportunities lost due to the inaccuracy of the product against the target value [10]. Taguchi creates a loss function as an equation:

$$L(y) = kv^2$$

Information:

v^2 = the standard deviation of the target value of the height of the anode

k $= \frac{A}{\Delta^2}$

A = the amount of loss due to the deviation is the average AVV used

Δ^2 = predetermined specifications

3.2 Operator Work Method Movement

In order for the pot voltage to remain stable, the anode replacement must be adjusted, every day 1 anode can be replaced. The operator movement will be described as a movement element based on the Modapts method. Work movements carried out by operators can be seen in Table 1.

Based on Table 1, basically, the body movements in this method are Therblig movements which are refined to become more detailed.

Table 1. Recapitulation of Operator Work Movements

No	Work	Movement Elements	Movement of Modapts
1	*ANODA CHANGING*	Turning on *crane*	*Finger, hand, foreamn, whole arm, extended arm, Grasp, press, eye regrasp*
2		Bring the crane to the pot area	*Finger, hand, foreamn, whole arm, extended arm, Grasp, press, eye regrasp*
3		Gripping the anode to be replaced	*Finger, hand, foreamn, whole arm, extended arm, Grasp, press, eye regrasp*
4		Removed the old anode	*Finger, hand, foreamn, whole arm, extended arm, Grasp, press, eye regrasp*
5		Observe the old anode	*Stand, eye*
6		Bring the old anode out of the pot	*Finger, hand, foreamn, whole arm, extended arm, Grasp, press, eye regrasp*
7		Place the anode above the comparison	*Finger, hand, foreamn, whole arm, extended arm, Grasp, press, eye regrasp*
8		Put it on the pallet	*Finger, hand, foreamn, whole arm, extended arm, Grasp, press, eye regrasp*
9		Check the state of the pot	*Stand, eye*
10		Place the new anode above the comparison	*Finger, hand, foreamn, whole arm, extended arm, Grasp, press, eye regrasp*
11		Bring the anode to the place that is replaced	*Finger, hand, foreamn, whole arm, extended arm, Grasp, press, eye regrasp*
12		Setting up a new anode	*Finger, hand, foreamn, whole arm, extended arm, Grasp, press, eye regrasp*
13		Bring the crane out of the pot and switch to another pot	*Finger, hand, foreamn, whole arm, extended arm, Grasp, press, eye regrasp*

3.3. Improvement of Operator Work Movement

Based on the economic principles of the movement, the worker's movement of the operator must fulfill the principles contained in the movement economy. The operator's work movements that are not in accordance with the principles of movement economics are shown in Table 2.

3.4. Processing Time

Calculation of process time is done by the Modular Arrangement of Predermined Time Standards (MODAPTS) method. In the actual working method, operators often make mistakes when replacing anodes, such as repetition of movements, workload imbalances between the left and right hands, and other operator errors, operators should be able to use idle time to maximize work movements when adjusting the height of the anode. Proposed improvements are made by reducing work (elimination) or repetitive movements and idle movements of all operators.

Table 2. Incompatible Operator Work Moves

No	Principles of Movement Economics	Operator Movement
1	Both hands must start and end their movements at the same time.	The operator's hand does not start and end the movement at the same time.
2	The movement of the body is saved, which is only moving the body parts that are needed to do the best work possible.	Operators use 42.4% and 38.6% of their body in the anode setting process.
3	Avoid movements that cause changes in direction because they will spend more time	There is no movement that causes changes in direction in the anode setting process
4	Place materials and work facilities in an easy and fast place to reach	Operator room is to be limited because of the space is very small, it is causing operator limited to make a movement when perform their task.

4 Conclusion

One particular causative factor that causes variation in the method of work carried out by the operator which later will directly affect the quality of the process. Based on this study, the Taguchi loss function method is used to calculate the amount of loss experienced by the company due to the variation by seeing the amount of power released during the activity. The total losses suffered by the company in Block 1 amounted to Rp 19,733,263, while the total loss in Block 2 was Rp 35,919,435. The results of the analysis of the operator's work method, namely the operator's movements are not in accordance with the economic principles of movement, namely the operator's hands do not start and end the movement at the same time, operator movement that should not be done when adjusting the new anode height, and operator space within the crane is limited because the area is too narrow. Comparison of the standard time of actual and proposed work methods is 774,847 seconds and 648.2 seconds with a time difference of 126,647 seconds faster and there is no movement that should not be done by the operator so the operator can use the time to adjust the anode height as well as possible according to the standards has been determined by the company.

References

1 S Assauri 2004 *Manajemen Produksi dan Operasi* (Jakarta: Lembaga Penerbit Fakultas Ekonomi Universitas Indonesia)
2 Siregar, I., Ginting, R., & Agnesia, F. (2017). Redesign Spring Bed Based on the Needs of Consumers. In *MATEC Web of Conferences* (Vol. 104, p. 03003). EDP Sciences
3 T Hani Handoko 2001 *Manajemen Personalia dan Sumber Daya Manusia* (Yogyakarta: BPFE Yogyakarta)
4 V Gaspersz 1998 *Statistical Processing Control : Penerapan Teknik-Teknik Statistical Dalam Manajemen Bisnis Total* (Jakarta: PT Gramedia Pustaka Utama)
5 Siregar, I., Nasution, A. A., & Sari, R. M. (2017,March). Effect of Total Quality Management on the Quality and Productivity of Human Resources. In *IOP Conference Series: Materials Science and Engineering* (Vol. 180, No. 1, p. 012116). IOP Publishing.
6 N Belavendram 1995 *Quality by Design: Taguchi Techniques for Industrial Experimentation* (London: Prentice Hall International)
7 Siregar, I., Suantio, H., Hanifiah, Y., Muchtar, M. A., & Nasution, T. H. (2017, January). Fuzzy linear programming for bulb production. In *Journal of Physics: Conference Series* (Vol. 801, No. 1, p. 012001). IOP Publishing.
8 I Soejanto *2008 Rekayasa Kualitas: Eksperimen dengan Teknik Taguchi* (Surabaya: Yayasan Humaniora)
9 Lawrence 2000 *Work Measurement and Methods Improvement* (Willey: New York)
10 J Antony 2001 *International Journal Advanced Manufacturing Technology*, Vol. 17
11 I Šabarić., S Brnada., Kovačević., Stana 2013 *Application of The MODAPTS Method with Innovative Solutions in The Warping Process*
12 S Sinulingga 2011 *Metode Penelitian* (Medan: USU Press)
13 G Taguchi 1989 *Introduction to Quality Engineering* (Asian Productivity Organisation)

Optimization of DND Multi-Depot Split-Load Pickup-Delivery Problem

Abdeslem Boukhtouta[1,a], Sujoy Ray[2], Andrei Soeanu[2], Raman Pall[1], Jean Berger[3]

[1]*Defence Research and Development Canada, Centre for Operational Research and Analysis, Ottawa, Ontario, Canada.*
[2]*International Safety Research Inc., 38 Colonnade Rd, Nepean, Ontario, Canada. K2E 7J6.*
[3]*Defence Research and Development Canada-Valcartier, Quebec, QC, Canada.*

Abstract. This paper presents a solution approach to optimize vehicle routes for a multi-depot, multi-vehicle, pickup and delivery problem over a large ground transportation network. More precisely, we address ground transportation of orders for the Canadian Department of National Defence using heterogeneous vehicle fleets. The fleets consist of limited number of organizational vehicles hosted at pre-established depots and commercial order delivery services. The proposed approach involves leveraging an insertion cost gradient-descent heuristic followed by a greedy randomized adaptive search procedure. Experimental results generated using the historical orders of the organization indicate that the developed approach is effective in handling a wide range of scenarios and may generate near-optimal vehicle routes with an annual transportation cost reduction between 7.7% and 16.7%.

1 Introduction

The Canadian Department of National Defence (DND) is using a heterogeneous fleet of vehicles for the ground transportation of goods between various bases and depots requesting orders through a central system. An order is an amount (weight) requested for delivery from a location to another location, at a specific date with a given priority. The request processing system is depicted in Figure 1. The system is responsible to develop routes and allocate orders to deliver in time using DND's own vehicles and commercial services. The DND's fleet of vehicles is known as Green Fleet (GF), which includes 5-Ton Small trucks (ST) and larger Tractor-trailers (TT). These are in limited in numbers and only a subset of locations, namely depots, host the GF vehicles. The route of an employed GF vehicle must start and end at the same depot where it is hosted. Moreover, each GF route in a solution must be scheduled at weekly basis for the whole year.

Thus, a year-long schedule for a GF vehicle is cost effective if the vehicle carries consignments on a regular basis. As such, contracted commercial fleet vehicles, also known as CF, supplement the GF to offer timely delivery of orders to various locations not served by the GF.

The analysis of historical data over DND's transport network revealed a critical need to optimize the problem of route planning for GF vehicles and efficiently using CF services to minimize the annual order delivery cost. In this paper, the resulting optimization problem is treated as an extension of the classical Pickup and Delivery Problem (PDP) under overarching organizational constraints. Thus, we define the problem of Multi-Depot Pickup-Delivery Problem with Split Loads (MDPDPSL) as follows.

Given a set of nodes (N) and a set of edges (E) where E is a relation of ($N \times N$), the transport network is a complete graph $G = (N, E)$. Each edge $\langle i, j \rangle \in E$ in the graph is a direct arc of distance d_{ij} between nodes i and j.

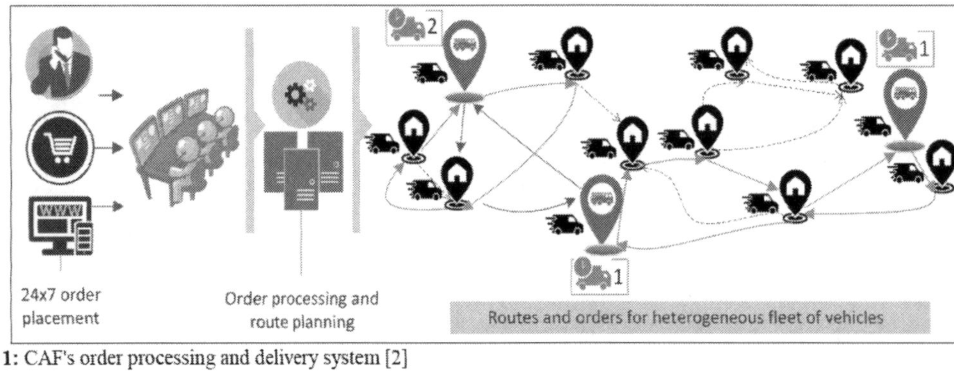

Figure 1: CAF's order processing and delivery system [2]

[a] Corresponding author: abdeslem.boukhtouta@forces.gc.ca

https://doi.org/10.1051/matecconf/201929602009

All nodes can employ unlimited commercial vehicles with a defined service cost per kg per km (*cf*). A subset of these nodes is selected as depots ($D, D \subseteq N$), where each depot can host a small number of GF vehicles. Each GF vehicle *k* has defined carriage capacity (Q_k) and service cost (gf_k). The processing system receives continuous orders where each order p is identified by a fixed amount q_{ij}^{p}, to be delivered from sender *i* to receiver *j*. An order can also be split and delivered by multiple vehicles, if needed. The aim of the problem is to deliver all orders at minimum total routing cost. The main objectives of this paper are:

- Analyze exact requirements and constraints for order pickups and deliveries among DND locations;
- Devise an automated approach to generate near-optimal solutions employing CF and GF fleets, using reasonable computing resources (time and memory);
- Evaluate the proposed approach using historical data from DND ground transportation network.

Manual order management over large transport network is often tedious, costly, error-prone and lacking process visibility. It also suffers from information and delays [1]. As such, end-to-end automation in order processing and delivery yields an improved supply chain quality control, efficient fleet use and notable cost reduction. The collected information allows better services and enhanced strategic decision making by applying state-of-the-art analyses. Also, seamless flow of such valuable information to-and-from external partners can collaboratively improve the reliability and performance of the supply chain.

In the remainder of the paper, Section 2 presents the problem and details the benefits of collaborative use of Green Fleet and Commercial Services to improve order pickup and delivery. Section 3 highlights previous efforts on PDP and related research problems. Section 4 illustrates the proposed multi-phase approach. Thereafter, Section 5 discusses the results obtained over four scenarios using DND's order delivery historical data. Finally, Section 6 provides the concluding remarks.

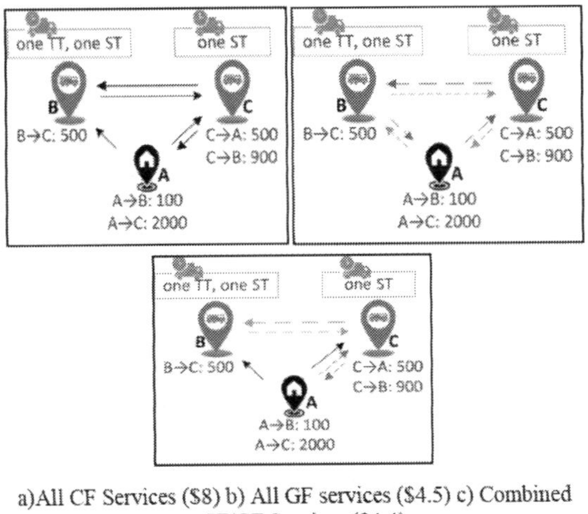

a)All CF Services ($8) b) All GF services ($4.5) c) Combined CF/GF Services ($4.4)

Figure 2: Comparison of solutions for the example [2]

2 Problem Description

Figure 2 shows the benefit of using different fleets from three locations A, B, C equidistant from each other by 100 km. The orders among A, B, C are: ‹A,B,100,D_1,1›, ‹A,C, 2000,D_1,1›, ‹B,C,500,D_1,1›, ‹C,A,500,D_1,1› and ‹C,B,900,D_1,1›. Each order has source, destination, weight (kg.), order date and priority. We consider the same date and the same priority for all orders. B hosts 2 GF vehicles: one ST (routing cost: 0.005$/km) and one TT (0.01$/km) with capacity 900 kg and 4000 kg respectively. C also hosts an ST vehicle. We evaluate solutions that minimize total routing cost for a commercial cost of 0.0002$/kg/km.

Figure 2(a) depicts the solution with commercial deliveries of all orders. The total routing cost is 8$. Figure 2(b) depicts a solution only using green vehicles efficiently where one TT is used from Node B on route: $B \rightarrow A \xrightarrow{2000} C \xrightarrow{900} B$ along with another ST vehicle from Node C on route: $C \xrightarrow{500} A \xrightarrow{100} B \xrightarrow{500} C$. Each vehicle respects its carriage capacity and yields total routing cost 4.5$. However, the solution cost can be further improved to 4.4$ using green and commercial fleet together as shown in Figure 2(c). The proposed GF routes are: $B \xrightarrow{500} C \xrightarrow{900} B$ and $C \xrightarrow{500} A \xrightarrow{900} C$. CF vehicles are used for additional shipment: $A \xrightarrow{100} B$ and $A \xrightarrow{1100} C$, marked by black arrows.

In practice, orders are grouped as per the schedules of the GF vehicles. Moreover, large orders are split in parts according to vehicles' availability and order priority. For example, two orders ‹A,B,100, 22/1/2018,1› and ‹A,B,2000, 23/1/2018,2› are placed to the processing system on Monday and Tuesday. Priority 1 and 2 require them to be delivered within 1 and 2 week(s) respectively. If the vehicles are scheduled on Monday and Thursday between A and B, the first order should be shipped on next Thursday while the second one can be shipped in next two consecutive schedules. Thus, on Thursday 25/1/2018, total combined transhipment between A and B will be $A \xrightarrow{1100} B$.

Figure 3 depicts our five-step strategy for order grouping. First, sensitive cargo orders are removed since they cannot be delivered by commercial services. Then, each order is divided into multiple transhipments as per vehicle availability and order priority. Statistical methods are used to eliminate unexpectedly high or low orders (outliers) since they can be served using commercial services. It is not beneficial to have schedules accommodating spurious raise in orders. Finally, an expected weekly amount is determined for every source and destination pair using one standard deviation over the average order weight.

Figure 3: Information pre-processing and transhipment splitting

3 Literature Review

Pickup and Delivery Problem (PDP) and its variants are research problems with NP-Hard complexity. Nowak et al. [3] studied the impact of order splitting over single-depot

PDP. Dividing an order over multiple vehicles needs less vehicles and saves transportation cost. However, according to our knowledge, impact of load splitting in multi-depot setting is very little assessed. Only Ríos-Mercado et al. [4] discussed a specific PDP routing problem in multi-depot and split-delivery setting.

Vehicle Routing Problem (VRP) is a simplified PDP where pickup or delivery location is same for all orders [5]. In more general case, PDP has no restrictions on rearranging commodity in vehicles and no fixed location for pickups or deliveries [6,5]. PDP is studied in relation to VRP [7] for goods deliveries, mail service, passenger transport, etc. Dial-A-Ride Problem (DARP) is a form of PDP where customers call for a ride from one location to another. DARP uses constraints on vehicle capacity and delivery time-window [8]. In practice, additional constraints are also used. VRP with Backhauls (VRPB) places constraints such that all deliveries precede the pickups to avoid rearrangement of goods in vehicles. For patient transportation, VRP with LIFO mandates that every picked-up patient is dropped off to a hospital before another pickup [9]. A multi-trip PDP with time-windows and manpower planning is detailed in [10] where non-emergency ambulance transfer is discussed. Analytic techniques (e.g. branch-and-cut and branch-and-price) are also used for PDP and DARP [11]. In addition, heuristic algorithms for single and multi-depot VRP with pickup and deliveries (MDPDP) are employed in [7].

Tabu search is applied in [9] on VRPPD with LIFO. A hybrid metaheuristic using genetic algorithm is detailed in [12] for VRP with multiple pickups, single delivery and time windows (VRPMPDTW). A genetic algorithm for multi-vehicle and multi-depot pickup and delivery with time windows is detailed in [13]. In [14], the authors employed an approach, leveraging iterative modified simulated annealing with greediness, for VRPPD. Particle Swarm Optimization is applied in [15] for pickup and delivery with time-windows (PDPTW). MDPDP with split load (MDPDPSL) is addressed in [4] using a Greedy Randomized Adaptive Search Procedure (GRASP). However, greedy algorithms have limitations for large problems [16]. In our case, this challenge is addressed by using a quick heuristic first to find good order delivery routes from each location. Then, the routes are extended by inserting other nearby locations. Finally, the routes are efficiently mixed using a GRASP based metaheuristic.

4 Proposed Approach

Given the scale of DND ground distribution network and complex operational constraints, optimal solution search was challenging. Unlike classical VRP, in MDPDPSL, each node can be a point of order pickup and/or delivery for any other nodes. Analytical approaches, such as branch and bound, branch and cut, etc., were inadequate to handle instances of practical size and failed to provide solution in reasonable time due to large number of decision variables and nonlinear nature of the problem. Thus, we use Heuristic approach to intelligently construct feasible solutions that are iteratively enhanced and compared. To improve solution quality, geographic features of the locations are exploited. An initial route is often extended to include additional routing points close to its path. This allows exploiting more opportunities for pickup and delivery among visited nodes. GF routes, computed using expected order amount, must also be maintained for a whole year. Only appropriate orders can be assigned periodically to the scheduled vehicles of limited capacities.

4.1 Problem Decomposition and Route Extension

First, we decompose the core problem into instances of a Split-Delivery VRP (SDVRP) variant for each order pickup location and solve each instance near-optimally using heuristics. In SDVRP, orders are picked from one location and served to various locations as per delivery needs. However, unlike classical SDVRP, in our case, GF vehicle(s) can be called from some of $D (D \subseteq N)$ depot(s) if the pickup location does not host adequate GF vehicles. Also, we allow split-delivery using heterogeneous fleets and make use of the availability of unlimited CF vehicles.

We extend an existing insertion cost gradient descent route construction heuristic. The technique initially computes an upper bound solution quickly as a reference using a nearest neighbour search. Then multi-tours are incrementally constructed by growing a pseudo-randomly selected tour with a nearby unserved node at a time. Two different supervision features guide the solution search: First, each multi-tour is stored in a sorted queue of fixed size in the ascending order of its total tour cost. Therein, a multi-tour of lowest cost is grown in each iteration. The fixed queue size also limits memory use and removes disqualified high cost multi-tours. Second, a map structure is maintained to record the serving costs for different fractions of total order serving. The map guides the solution search by offering early hint on whether a growing multi-tour has potential to produce a near-optimal solution. When a multi-tour serves all the nodes, it forms a solution. The lowest-cost solution is always retained and any multi-tour exceeding that cost gets eliminated from the sorted queue. The obtained solutions of all individual SDVRP instances are then combined to a final solution with vehicles starting from multiple depots.

In the route extension phase, each vehicle route is extended to add additional locations close to the route. The underlying method first finds nearest neighbours for each leg of a route within a cost threshold. Then, it adds a neighbour node into the route between two nodes if and only if the new routing cost through the inserted node does not exceed a fixed small percentage of the original routing cost between the two nodes. This allows efficient loading of GF vehicles with more orders between the nodes of the route where the pickup nodes are different from the original supply node of an individual vehicle path.

4.2 GRASP and Vehicle Route Selection

The GRASP step improves the combined solution by shifting order(s) from one vehicle to other when the latter passes through the same order pickup and delivery locations in sequence and has enough remaining cargo capacity. Efficient order loading helps reducing transport cost by eliminating redundant and inefficient CF and GF vehicles from the original solution. We iteratively perform pseudo-random order selection with a bias to its cost benefit ratio. GRASP depends on two strategies: Diversification and Intensification. First one involves a *LocalSearch* method to approximately evaluate cost benefits by transferring the selected order to available vehicles and determines the suitable one. Transfer of the order forms a new solution. Intensification is implemented through *PathRelinking* which improves both, the current best and the newly found solutions by exchanging their distinct features. Different order delivery strategies of previously found solutions are also analysed. This may lower the cost of both solutions. After each iteration, the current best solution is retained.

The cost of each route is further evaluated against the option of delivering associated orders using CF service. Then most efficient GF routes are selected from the depots, as per vehicle availability and vehicle reuse constraints. The remaining orders are then delivered commercially.

4.3 Benefits and Drawbacks

The proposed approach allowed handling large-scale problem instances of order deliveries in multiple phases. Decomposing MDPDPSL into SDVRP instances helped treating the main problem concurrently and with lower memory footprint per sub-problem. Due to unlimited CF vehicle availability, a valid solution is always guaranteed for any of our problem instances. Geographic positions of pickup and delivery locations also helped in cost saving by route extension. However, initial SDVRP routes are formed using recipients' locations for every sender. This reflects a bias to produce cost efficient routes from those locations sending large amount of orders but an initial subset of less efficient routes, computed from SDVRP instances, could be found highly efficient once more orders are assigned between nodes of the route. Moreover, lack of processing ability limits evaluation of route extension phase within a distance threshold from the initial route. Also, greedy approach (GRASP) to improve vehicle's delivery efficiency is fast but not always optimal.

5 Results and analysis

We tested the approach on four years historical order delivery data, across 308 DND locations including 14 depots. These locations are operational from Monday to Friday between 7:30am and 4pm. Unlimited CF and limited numbers of GF vehicles (ST and TT) are used in four scenarios. Each GF vehicle is scheduled to start from its depot on every Monday and/or Thursday of the week to maximize operation time of the locations. Vehicles returning to depot in two or three days are often reused for next weekly schedule. The travel costs of ST and TT are respectively \$0.65 and \$1.00 per km Commercial fleet charges \$0.000229 per km per kg in each scenario except the last one where an advanced cost estimation technique is used (will be described later). The time to unload and load orders at any base is considered as 1 hour. A full day off is also added for each driver's schedule after 5 days of work as per regulations. In this setting, the purpose of the scenarios are as follows:

- Scenario 1 minimizes total order delivery cost without any restriction on the selection of depots among all bases and availability of CF and two types of GF fleets.
- Scenario2 estimates the cost and route changes over Scenario 1 by closing Toronto (ON) base and shifting all its orders to Angus-Borden (ON) base in proximity.
- Scenario 3 aims to minimize the order delivery cost by specifying candidate depots and availability of green vehicles only from these depots: Angus-Borden (ON), Bagotville (QC), Cold Lake (AB), Comox (BC), Edmonton (AB), Esquimalt (BC), Gagetown (NB), Halifax (NS), Greenwood (NS), Montreal (QC) and Valcartier (QC). Among these locations, Comox hosts 1ST, Montreal hosts 4 TTs and 2 STs. Each of the other depots hosts only 1 TT. Furthermore, each ST route is restricted to a tour duration spanning no more than 2 days while adding \$150 for each overnight stay of ST truck. GF drivers are allowed up to 13 hours of driving within their daily schedule of 14 work hours.
- Scenario 4 assesses near optimal order delivery cost on the same constraints of Scenario 3 but alters the linear estimated rate of CF service with a more realistic market driven cost estimation method.

The estimation leverages a data under-sampling-technique that exploits k-nearest neighbours (k-NN) algorithm over historical data to predict the transport cost. First, Density Based Scanning (DBScan) finds a cluster of at least k closest data points with their relative distances from the input order data in a multidimensional space. The scanning stops when k neighbours are found in its largest cluster. The measured distance of these data points from input helps computing the order delivery cost. We notice that linear estimation is reasonably useful for orders weighing up to 2000 kg., as the carriers compete to deliver the orders. However, for orders with weights over 2000 kg., linear estimation is less accurate. The proposed method is useful for the entire range of weights, particularly for large weights. Figure 4 depicts total number of depots and GF vehicle tours (certain vehicles are reused) for every week in each of the scenarios. Table 1 details the experimental results. Four years of selected order deliveries actually costed DND \$34.5 million. The cost would increase to \$45 million if only CF services would be used at a linear cost of \$0.000229 per km per kg. However, using combined GF and CF services, significant saving can be achieved. Scenario 4 indicates a better CF service cost estimation technique (deviation: ±5% of actual service cost) may save 13.41% to 16.75% of operation cost.

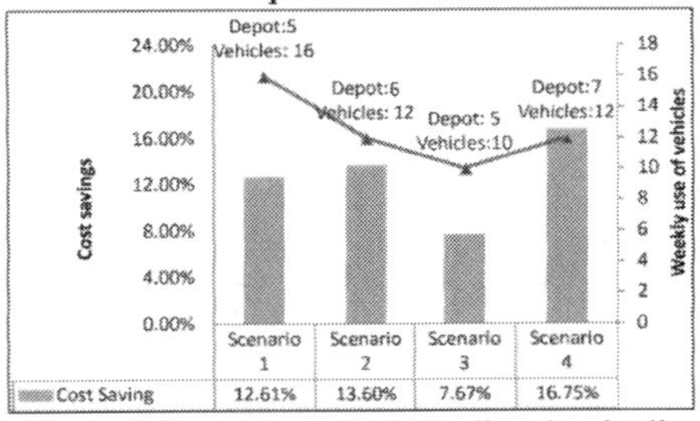

Figure 4: Cost saving and vehicle allocation details

Table 1: Experiments on four years DND order delivery data

	GF Cost ($)	CF Cost ($)	Total Cost ($)	Savings (%)
Scenario 1				
Estimated all orders served commercially (Linear)	0	45,023,417	**45,023,417**	
Estimated current operational cost	7,546,240	40,787,693	**48,333,933**	
Actual current operational cost (database)	7,546,240	26,968,443	**34,514,683**	
Estimated proposed operational cost: 1std (Linear)	8,295,785	21,866,053	**30,161,838**	**12.61**
Scenario 2				
Estimated all orders served commercially (Linear)	0	44,961,180	**44,961,180**	
Estimated current operational cost	7,635,680	40,753,909	**48,389,589**	
Actual current operational cost (database)	7,546,240	26,968,443	**34,514,683**	
Estimated proposed operational cost:1std (Linear)	6,010,020	23,808,863	**29,818,883**	**13.60**
Scenario 3				
Actual current operational cost with new constraints	7,546,240	26,968,443	**34,514,683**	
Estimated proposed operational cost (1std)	7,925,069	23,940,814	**31,865,883**	**7.67**
Scenario 4				
Estimated all orders served commercially (Clustering)	0	32,312,903	**32,312,903**	
Estimated current operational cost (Clustering)	7,546,240	27,546,047	**35,092,287**	
With commercial service overestimate correction (95%)			**33,714,985**	
With commercial service underestimate correction (105%)			**36,469,589**	
Estimated proposed operational cost (1std) - (Clustering)	5,675,524	23,057,140	**28,732,664**	**16.75**
With commercial service overestimate correction (95%)			**27,579,807**	**20.09**
With commercial service underestimate correction (105%)			**29,885,521**	**13.41**

6 Conclusions and future work

The proposed approach employs initial route generation and subsequent metaheuristic route improvement, yielding competitive solutions for relevant size PDP problems.

We examined four scenarios of growing difficulty in terms of depot locations, vehicle availability, routing constraints and cost estimation. The use of heterogeneous vehicles along with split loading provides notable cost savings due to improved capacity use in terms of cargo weight. Cargo volume is subject of future work though. The approach is suitable for transport management, aiming at cost effective order delivery by different types of vehicle with specific trade-offs in terms of routing cost versus capacity or loading level. As such, the approach is anticipated to allow DND to carry out mission planning with reduced distribution cost and improved sustainment.

References

1. Esker, Improving Customer Service and Fulfillment with Order Processing Automation, online, https://quitpaper.esker.com/Gartner-Improving-Customer-Service-and-Fulfillment-document-library.html (March 2019)
2. Boukhtouta A. et al., Modelling and Optimizing DND Ground Transportation Distribution Network, DRDC Scientific Report, 83 pages, (2019).
3. M. Nowak, A. Ergun, C. C. White, Pickup and Delivery with Split Loads, *Transportation Science* **42**, 32–43, (2008).
4. R. Z. Ríos-Mercado, J. F. López-Pérez, A. Castrillón-Escobar, A GRASP for a Multi-depot Multi-commodity Pickup and Delivery Problem with Time Windows and Heterogeneous Fleet in the Bottled Beverage Industry, *Springer*,143–157, (2013).
5. M.W. P. Savelsbergh, M. Sol, The general pickup and delivery problem, *Transp. Science* **29**, 17–29, (1995).
6. G. Berbeglia, J-F. Cordeau, I. Gribkovskaia, G. Laporte, Static pickup and delivery problems: A classification scheme and survey, *TOP* **15**, 1–31, (2007).
7. G. Nagy, S. Salhi, Heuristic algorithms for single and multiple depot vehicle routing problems with pickups and deliveries, *EJOR* **162**, 126–141, (2005).
8. J-F. Cordeau, G. Laporte, The dial-a-ride problem: models and algorithms, *Annals of Operations Research* **153**, 29–46, (2007).
9. E. Benavent, M. Landete, E. Mota, G. Tirado, The multiple vehicle pickup and delivery problem with LIFO constraints, *EJOR* **243**, 752 – 762, (2015).
10. A. Lim, Z. Zhang, H. Qin, Pickup and Delivery Service with Manpower Planning in Hong Kong Public Hospitals, *Trans. Science* **51,** 688-705, (2016).
11. L. Xue, Z. Luo, A. Lim, Exact approaches for the pickup and delivery problem with loading cost, *Omega* 59, Part B, 131 – 145, (2016).
12. S. Yanik, B. Bozkaya, et al., A new VRPPD model and a hybrid heuristic solution approach for e-tailing, *EJOR* **236**, 879–890, (2014).
13. E. B. Alaia, I. H. Dridi, H. Bouchriha, P. Borne, Optimization of the multi-depot & Multi-vehicle pickup and delivery problem with time windows using genetic algorithm, *IEEE-CoDIT*, 343–348, (2013).
14. G. Martinovic, I. Aleksi, A. Baumgartner, Single-commodity vehicle routing problem with pickup and delivery service, *Mathematical Problems in Engineering* (2008).
15. P. Sombuntham, V. Kachitvichyanukul, Multi-depot Vehicle Routing Problem with Pickup and Delivery Requests, *American Institute of Physics Conf. Series*, **1285**, 71–85, (2010).
16. M. G. C. Resende, C. C. Ribeiro, Greedy Randomized Adaptive Search Procedures, Springer, 219–249, (2003).

AUTHOR INDEX